COLOUR AND OPTICAL PROPERTIES OF MATERIALS

An Exploration of the Relationship
Between Light, the Optical Properties
of Materials and Colour

Richard J. D. Tilley
Cardiff University

JOHN WILEY & SONS, LTD
Chichester • New York • Weinheim • Brisbane • Singapore • Toronto

OTHER WILEY EDITORIAL OFFICES

John Wiley & Sons, Inc., 605 Third Avenue, New York, NY 10158-0012, USA

WILEY-VCH Verlag GmbH, Pappelallee 3, D-69469 Weinheim, Germany

Jacaranda Wiley Ltd, 33 Park Road, Milton, Queensland 4064, Australia

John Wiley & Sons (Asia) Pte Ltd, 2 Clementi Loop #02-01, Jin Xing Distripark, Singapore 129809

John Wiley & Sons (Canada) Ltd, 22 Worcester Road, Rexdale, Ontario M9W 1L1, Canada

LIBRARY OF CONGRESS CATALOGING-IN-PUBLICATION DATA

Tilley, Richard J. D.
   Colour and optical properties of materials : an exploration of the
relationship between light, the optical properties [i.e. properties]
of materials and colour / Richard J.D. Tilley.
      p. cm.
   Includes bibliographical references and index.
   ISBN 0-471-85197-3 (alk. paper). – ISBN 0-471-85198-1 (pbk. :
alk. paper)
   1. Light. 2. Optics. 3. Colour. I. Title.
QC355.2.T55 1999
535.6–dc21   99–35920
                                                                    CIP

BRITISH LIBRARY CATALOGUING IN PUBLICATION DATA

A catalogue record for this book is available from the British Library

ISBN 0-471-85197-3 (cloth)
ISBN 0-471-85198-1 (pbk)

Typeset in 10/13pt Caslon 224 by Mayhew Typesetting, Rhayader, Powys
Printed and bound in Great Britain by Biddles Ltd, Guildford and King's Lynn

This book is printed on acid-free paper responsibly manufactured from sustainable forestry, in which at
least two trees are planted for each one used for paper production.

Colour and Optical Properties of Materials

CONTENTS

x
−

PREFACE

This book is concerned with colour. The aim is to give a flavour of the many ways that colour can be produced and be put to use in society. As it is not possible to discuss colour without reference to numerous other optical properties, these too are explained throughout the text. Colour, however, remains the dominant theme.

The organisation of the book falls into three recognisable parts. Chapters 1 to 6 are concerned with the interaction of light with transparent materials. Chapters 7 to 10 develop the way that light interacts at an atomic scale to produce colours. Finally Chapters 11 to 13 outline applications that are of importance in everyday life. None of this coverage is to be found in any single contemporary book. Indeed, many aspects of science and engineering are intertwined and the contents touch upon Physics, Chemistry, Biology, Materials Science and Engineering, and on Electrical and Electronic Engineering. Students of all of these disciplines should find this book of relevance to their studies or interests.

Readers who need more information can turn to the Further Reading sections at the end of each chapter, which include selected references to the original literature or substantial reviews and will allow them to take matters further. In addition, each chapter contains some exercises and numerical problems which have been provided to illustrate and reinforce the concepts presented in the text. All readers are encouraged to attempt them. Full annotated solutions to these problems are to be found on the World Wide Web (see p. 317).

It is a pleasure to acknowledge the considerable help and encouragement received in the preparation of this book. Elizabeth, Gareth and Richard Tilley all read early versions of the text and gave continual encouragement. Professor D. B. Holt of Imperial College of Science, Technology & Medicine, Professor I. J. McColm of the University of Bradford, Professor F. S. Stone of the University of Bath and Dr A. Towns, University of Leeds read the whole of the draft manuscript and provided an enormous number of sage and helpful comments which have materially added to the scope and perspective of the book. Dr A. Slade of John Wiley has always given both assistance and encouragement in the venture. To all of these I express my sincere thanks. I also would like to record my indebtedness to Dr J. A. Findlay, who provided

Plate 4.3a, Spectrum Technologies plc, Bridgend, for Plate 9.1 and Dr A. Eddington, who provided Plate 9.2b. I should also like to express my thanks to Dr R.M. Lloyd, who created the web site containing the annotated solutions to all the problems and exercises.

Finally, my thanks, as always, are due my wife Anne, who tolerated my hours reading or sat in front of a computer without complaint, and made it possible to complete this work.

CHAPTER 1

LIGHT AND COLOUR

What is light?
What is colour?
Why do hot objects become red or white hot?
How do colour filters affect colour?

1.1 LIGHT

Light has been a puzzle from earliest times and remains so today. In elementary optics light can usefully be considered to consist of light *rays* and the majority of optical instruments can be constructed within the framework of this idea. However, the ray concept breaks down when the behaviour of light is critically tested. At this stage more complex ideas are needed.

The first testable theory of the nature of light was put forward by Newton, who suggested that it was composed of small particles or "corpuscles". This idea was supported on philosophical grounds by Descartes. Huygens, a contemporary, thought that light was wavelike, a point of view also supported by Hooke. Over the years the wave theory gradually came to take precedence, and was strengthened by the theoretical work of physicists such as Fresnel, who first explained interference (p. 6) and diffraction (Chapter 6) using wave theory. Polarisation (Chapter 3) is similarly well explained on the assumption that light is a wave.

The two theories differed in one fundamental aspect that could be tested. When light enters water it is refracted (see Chapter 2). In terms of corpuscles, this implied a speeding up of the light in water relative to air. The wave theory demanded that the light should move more slowly in water than air. The experiments were complicated by the enormous speed of light, which was known to be about 3×10^8 m s^{-1}, and it was not until April 1850 that Foucault first proved that light moved slower in water than air, and seemingly killed the corpuscular theory then and there. Confirmation of the result by Fizeau a few months later removed all doubt.

The wave theory of light undoubtedly reached its peak when Maxwell developed his theory of electromagnetic radiation and showed that light was only a small part of an *electromagnetic spectrum*. Maxwell's theory was confirmed experimentally by Hertz whose experiments led directly to radio.

The problem for the wave theory was that waves had to exist in something, and the "something" was hard to pin down. It became called the "luminiferous aether" and had the remarkable properties of pervading all space, being of very small (or even zero) density and having extremely high rigidity. Attempts to measure the velocity of the Earth relative to the luminiferous aether, the so-called aether drift, by Michelson and Morley, before the end of the 19th century, proved negative. The difficulty was removed by Einstein's theory of relativity and for a time it appeared that a theory of light as electromagnetic waves would finally explain all optical phenomena.

This proved a false hope and the corpuscular theory of light was revived early in the 20th century. Since 1895 it had been observed that when ultraviolet light was used to illuminate the surfaces of certain metals, negative particles, later identified as electrons, were emitted. The details of the experimental results were completely at odds with the wave theory. The electrons, called *photoelectrons*, were only observed if the frequency of the radiation exceeded a certain minimum value which varied from one material to another. The number of photoelectrons emitted increased as the intensity of the light increased, but their energy remained constant for any particular light source. Very dim illumination still produced small numbers of photoelectrons with the appropriate energy and it was found that the kinetic energy of the photoelectrons was linearly proportional to the frequency of the illumination.

The explanation of this "photoelectric effect" by Einstein was based upon the fact that light behaved as small particles, now called *photons*. He proposed that the light photons had an energy given by:

$$E = h\nu$$

where ν was the frequency of the radiation and h a constant, Planck's constant, which we will meet again below. The kinetic energy of the photoelectrons could then be written as:

$$\tfrac{1}{2}mv^2 = h\nu - \phi$$

where ϕ is known as the *work function* of the metal. Einstein thus rescued the wave theory from the dilemma of the luminiferous aether and then seemingly wrecked the selfsame theory via his explanation of the photoelectric effect.

At present all experiments show that light is best described as consisting of streams of photons. However, the statistical behaviour of a large number of photons is represented very well by an electromagnetic wave.

In this book we will try to avoid the confusion that this can sometimes cause by giving explanations in terms of the simplest approach which is in accord with the observations. For some phenomena it is adequate to use the idea of a ray of light. When objects with dimensions of the same size as a light wave are encountered it is necessary to consider light to be a wave. Atomic processes often require a photon approach. It needs to be stressed that these are not different fundamentally. All are contained within the most advanced theory of optics available today, generally described as quantum optics or quantum electrodynamics. The way in which these mutually complementary approaches fit together is detailed in some of the sources listed in the Further Reading section.

1.2 LIGHT WAVES

Some aspects of the way in which light behaves are conveniently treated by considering light as part of the electromagnetic spectrum, as shown in Figure 1.1. That is, light is regarded as a wave of wavelength λ with an electrical and magnetic field, each described by a vector. As far as the topics in this book are concerned the magnetic field need not be considered and only the electric field needs to concern us. Light waves can then conveniently be depicted as in Figure 1.2.

This figure is represented by the equation:

$$y = a_0 \sin \left[(2\pi/\lambda) \, (x + vt) \right]$$

Here y is the magnitude of the electric field vector (which lies in the plane of the figure) at position x and time t and a_0 is the *amplitude* of the wave and is a constant. The peaks in the wave are referred to as *crests* and the valleys as *troughs*. Any point on the wave, a crest say, is moving in the x direction with a velocity v.

It is sometimes more convenient to represent the wave equation in terms of the frequency of the vibration of the wave. The velocity of the wave, v, is related to the frequency, ν, by the equation:

$$v = \lambda \nu$$

4

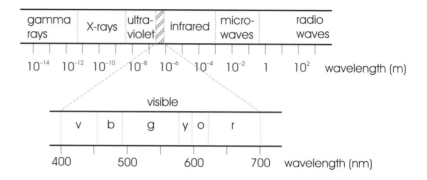

v, violet; b, blue; g, green; y, yellow; o, orange; r, red

Figure 1.1 The electromagnetic spectrum. Historically, different regions have been given different names. The boundaries between each region are not sharply defined but grade into one another. The visible spectrum occupies only a small part of the total spectrum

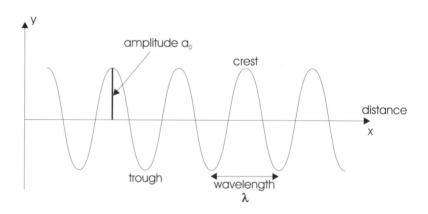

Figure 1.2 Part of a light wave travelling along x. The curve represents the magnitude of the electric field vector, y, as a function of position. The distance between the crests or troughs is the wavelength, λ. Any point on the wave moves with a velocity v. If the electric field vector remains in the plane of the paper, as drawn, the light is linearly polarised. If the orientation of the electric field with respect to the plane of the page varies at random so that the curve continually adopts differing angles with the plane of the paper, the light is unpolarised

The frequency, ν, has units of Hz (Hertz) or s^{-1}. The equation then becomes:

$$y = a_0 \sin \left[(2\pi/\lambda)\ (x + \lambda \nu t)\right]$$

Further mathematical simplification can be achieved if the substitutions:

$$k = 2\pi/\lambda$$

$$\omega = 2\pi\nu$$

are used, where k is the *wave number* (with units m^{-1}) and ω is the *angular frequency* (with units of radians s^{-1}). Making these substitutions, the wave equation becomes:

$$y = a_0 \sin (kx + \omega t)$$

Our eyes can detect only a small part of the whole electromagnetic spectrum, called the *visible spectrum*, as shown in Figure 1.1. Moreover, we detect the *intensity* of the wave rather than its amplitude. The intensity, I, is proportional to the square of the amplitude:

$$I = K(a_0)^2$$

where the value of the constant of proportionality, K, depends upon the properties of the medium containing the wave.

The extent of the visible spectrum is defined in terms of the wavelength or frequency of the light waves involved. *Perception* of the different wavelengths is called *colour*. The precise measurement of colour involves a determination of the energy present at each wavelength in the light using a spectrometer (see p. 147).

The shortest wavelength of light that we can perceive corresponds to the colour violet, with a wavelength near to $\lambda = 400$ nm. The longest wavelength of light perceived corresponds to the colour red, with a wavelength close to $\lambda = 700$ nm. Between these two limits the other colours of the spectrum occur in the sequence from red to orange, green, blue, indigo and finally to violet (see Figure 1.1 and Table 1.1).

Wavelengths shorter than violet fall in the *ultraviolet*. "Ultraviolet A" is closest to the violet region and "ultraviolet B and C" are at shorter wavelengths. Ultraviolet B and C radiation are able to damage biological cells and lead to sunburn and possibly the occurrence of skin cancer.

Table 1.1 The visible spectrum

Colour	λ (nm)	ν (Hz)	Energy (J)	Energy (eV)
Deep red	700	4.29×10^{14}	2.84×10^{-19}	1.77
Orange-red	650	4.62×10^{14}	3.06×10^{-19}	1.91
Orange	600	5.00×10^{14}	3.31×10^{-19}	2.06
Yellow	580	5.17×10^{14}	3.43×10^{-19}	2.14
Yellow-green	550	5.45×10^{14}	3.61×10^{-19}	2.25
Green	525	5.71×10^{14}	3.78×10^{-19}	2.36
Blue-green	500	6.00×10^{14}	3.98×10^{-19}	2.48
Blue	450	6.66×10^{14}	4.42×10^{-19}	2.75
Violet	400	7.50×10^{14}	4.97×10^{-19}	3.10

Radiation with wavelengths longer than red are referred to as *infrared* radiation. Although not visible, the longer wavelengths of infrared radiation, called *thermal infrared*, are detectable as the feeling of warmth on the skin.

A beam of light is said to be *monochromatic* when it is comprised of only a very narrow range of wavelengths and *coherent* when all of the waves which make up the beam are completely *in phase*, that is, the crests and troughs of the waves are in step. If the electric field vector shown in Figure 1.2 remains in one plane the light is said to be *linearly* (or *plane*) *polarised*.

Normal light is not emitted in a continuous stream, but in short bursts lasting about 10^{-8} s. Within each burst all of the light waves are in phase and linearly polarised. However, both the phase and polarisation change from burst to burst in a random fashion so that the phase and polarisation of each burst is unrelated to those coming before and after. Within the space of a fraction of a second the phase and the polarisation of a light wave fluctuates continuously and at random. Normal light is thus described as being *incoherent* and *unpolarised*.

1.3 INTERFERENCE

One of the advantages of the wave description of light is that interference phenomena are easily explained. The idea is illustrated in Figure 1.3. If two light waves occupy the same region of space at the same time they can add together, or *interfere* to form a product wave. This idea, called the *principle of superposition*, was stated by Young some two centuries ago, in 1802. If two identical waves are exactly in step then they will add to produce a

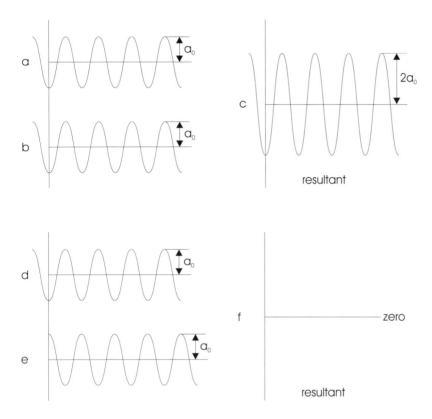

Figure 1.3 (a–c) The addition of two waves in phase (in step), (a), (b), will produce a wave of twice the amplitude of the original wave (c). (d–f) The addition of two waves out of phase (out of step) by λ/2, (d), (e), will produce a wave with zero amplitude (f)

resultant wave with twice the amplitude (Figure 1.3a–c) by the process of *constructive* interference. If the two waves are out of step the resultant amplitude will be less due to *destructive* interference. If the waves are sufficiently out of step that the crests of one correspond with the troughs of the other, the resulting amplitude will be zero (Figure 1.3d–f).

1.4 LIGHT PHOTONS

When dealing with events at an atomic scale it is often best to regard light as composed of particles, called photons. The energy of a photon is given by:

$$E = h\nu = hc / \lambda$$

where h is Planck's constant (6.626×10^{-34} J s) and c is the velocity of light in vacuum (2.998×10^8 m s^{-1}). The conjunction of the particle and wave descriptions, called wave-particle duality, is evident in the fact that ν is the frequency (s^{-1}) and λ is the wavelength (m) of the wave-like properties associated with the photon.

The relationship between the wavelength and the frequency is:

$$\nu\lambda = c$$

The colours of the spectrum as perceived by a normal eye, together with the appropriate values of wavelength, frequency and energy of the waves or photons are set out in Table 1.1. The divisions between the colours are, of course, artificial and each colour blends into its neighbours.

1.5 BLACK-BODY RADIATION AND INCANDESCENCE

Incandescence is the emission of light by a hot body. The sun and tungsten lamp filaments provide commonplace examples and both would be regarded as producing white light. The yellow light characterising the upper part of a candle flame also arises from incandescence. In this case small particles of carbon are heated to temperatures of 1200–1400°C and emit light which we perceive as more or less yellow in colour. When light from an incandescent object is spread out according to wavelength by a prism (see Chapter 2) the result is the continuous fan of colours listed in Table 1.1, known as a *continuous spectrum*.

Incandescence comes about in the following way. At absolute zero all atoms and molecules making up the solid are in the lowest possible energy state. As the temperature increases they absorb energy and are promoted to higher energy levels and, at the same time, atoms and molecules which have already absorbed energy emit energy as they fall back to lower energy levels. The energy levels involved in this process will be described in later chapters. For the moment it is only necessary to know that the radiation emitted effectively extends over a continuous range of wavelengths. For a solid a little above room temperature all the wavelengths of the emitted energy lie in the infrared and although the radiation is invisible it is detectable as a sensation of warmth. At a temperature of about 700°C the longest wavelengths emitted creep into the red end of the visible spectrum.

The colour of the emitter is seen as red and the object is said to become *red hot*. At higher temperatures the wavelengths of the radiation given out extend increasingly into the visible region and the colour observed changes from red to orange and thence to yellow, as in the example of a candle flame, mentioned above. When the temperature of the emitting object reaches about 2500°C all visible wavelengths are present and the body is said to be *white hot*, as in the case of an ordinary light bulb.

A *black body* is an idealised object which absorbs and emits all wavelengths perfectly. A small blackened sphere containing a pin-hole approximates to a black body. If the intensity of the radiation issuing from the pin-hole when such a sphere is heated is measured as a function of wavelength the *black-body emission spectrum*, shown in Figure 1.4, is obtained. The actual shape of the curve is found to be dependent only upon the temperature of the body. As the temperature increases the peak in the curve moves to shorter wavelengths (that is, higher energies).

The most important incandescent object for us is the sun, which is the ultimate source of energy on Earth. The solar spectrum has a form quite similar to a black body curve with a maximum near 560 nm. The curve corresponds to a solar temperature of about 5700°C (about 6000 K). Light is perceived as white if it has a make-up like that of the solar spectrum. The human eye is most sensitive to the maximum in the solar spectrum, which corresponds to yellow-green and is noticeably less sensitive to blue and red light (Section 1.6).

The successful theoretical prediction of the form of the emission curve in Figure 1.4 by Planck in 1901 signalled the start of the modern age of quantum theory. The formula derived by Planck to describe the energy density, $\rho(\lambda)d\lambda$, which is the energy emitted (in J m^{-3}) as a function of the wavelength λ by a black body at a temperature T was:

$$\rho(\lambda)d\lambda = 8\pi hc \ d\lambda \ / \ \lambda^5[\exp(hc/\lambda kT) - 1]$$

where h is Planck's constant (6.626×10^{-34} J s), c is the speed of light (2.998×10^8 m s^{-1}), λ is the wavelength (m), k is Boltzmann's constant (1.380×10^{-23} J K^{-1}) and T the temperature of the body (K). This equation is also commonly expressed in terms of the frequency of the radiation, ν, rather than the wavelength:

$$\rho(\nu)d\nu = 8\pi h\nu^3 \ d\nu \ / \ c^3 \ [\exp(h\nu/ kT) - 1]$$

The other symbols have the same meaning as above.

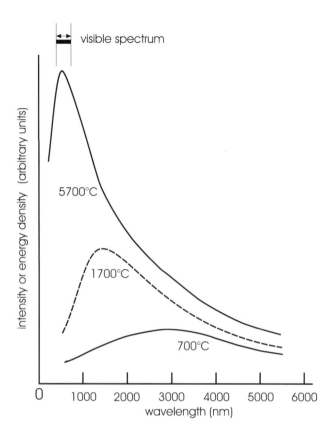

Figure 1.4 The intensity of radiation emitted from a black body as a function of wavelength. The maximum of the curve increases and moves towards shorter wavelengths (higher energies) at higher temperatures. The spectrum emitted by the sun is similar to that for a black body at 5700°C and that from a red-hot object is similar to the curve for a black body at 700°C. The maximum sensitivity of the human eye matches the peak in the solar spectrum

The revolutionary concept that Planck employed in the derivation of these equations to successfully reproduce the black-body curve was that the radiation was not emitted over a continuous spread of frequencies, but that only some frequencies were represented in the process. Moreover, the energy absorbed or given out by the atoms and molecules in the black body had to be delivered in packets or *quanta*. Energy emission was said to be *quantised*. The relationship between the energy of a single quantum, E, and the frequency of the radiation, ν, was given by what has since become one of the most famous equations of science:

$$E = h\nu$$

As we saw, this equation was exploited by Einstein in his explanation of the photoelectric effect.

1.6 THE COLOUR OF LIGHT: ADDITIVE COLORATION

Colour is not especially easy to define. The word refers to the physiological response of the eye–brain combination to light waves falling upon the light-sensitive *retina* which makes up the inner surface of the eye. The light receptors found in the retina are of two types, cone cells, which do not respond to colour but are sensitive to low light levels, and rod cells, which provide colour vision in bright light (see Chapter 8). The sensitivity of the eye to colour depends not only upon the light intensity, but also upon which area of the retina is being stimulated. The most sensitive region, called the *fovea*, is almost directly behind the lens of the eye. The sensitivity of a normal eye to bright white light focused on the fovea is drawn in Figure 1.5. The maximum sensitivity is for a wavelength close to 555 nm. This matches the peak intensity of sunlight at the surface of the Earth.

It appears that an average person can distinguish more than a million different colours. These are all the result of colour mixing, which can be due to *additive colour mixing* or *subtractive colour mixing*, both described below. The colours of the spectrum are called *chromatic* colours. Non-chromatic colours are those like brown, which do not appear in the spectrum.

Despite the complexity inherent in the concept of colour and its perception, it has been found that all colours can be precisely specified by just three parameters.

(i)   *Hue*, which corresponds to the wavelength or frequency of the radiation. The hue is given a colour name such as red or yellow.
(ii) *Saturation* or *chroma*, which corresponds to the amount of white light mixed in with the hue and allows pale "washed out" colours to be described.
(iii) *Lightness*, *brightness* or *value*, which describes the intensity of the colour, the number of photons reaching the eye.

Colours can then conveniently be represented in a three-dimensional coordinate system. The form most often chosen is cylindrical, with the hue

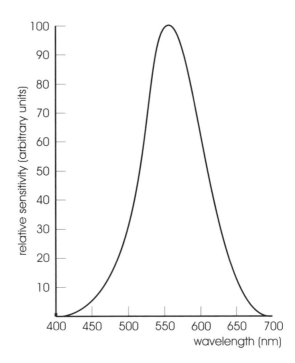

Figure 1.5 The sensitivity of a normal eye to bright white light focused on the fovea. The maximum sensitivity is for a wavelength close to 555 nm

being arranged around the periphery of the cylinder, the saturation being represented by the radius and the lightness by the cylinder axis, as illustrated in Figure 1.6a. This arrangement is called the *Munsell colour cylinder* or *Munsell colour solid*.

Additive colour mixing occurs when two or more beams of differently coloured light combine. It has been found that the majority of colours can be produced by mixing just three *additive primary colours*, red, green and blue. (Strictly speaking any monochromatic colours near to these colours will suffice.) In general a colour can be made up of certain quantities, called the *tristimulus values*, r of the *red* component, g of the *green* component and b of the *blue* component, thus:

$$colour = r + g + b$$

This is called the RGB colour model. The concept allows one to represent colours by a planar diagram. The first step is to draw the red, green and

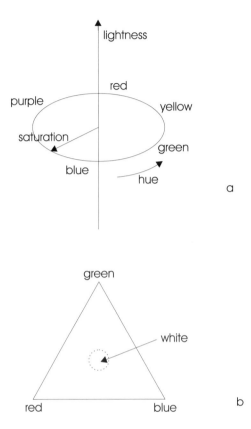

Figure 1.6 (a) The representation of colours on a cylindrical colour coordinate system. The hue is given by a point on the circumference of the cylindrical system, the saturation by the distance along the radius from the axis of the cylinder and the lightness by the vertical axis of the system. (b) A colour triangle. The hue and saturation of a colour can be represented by a point in the plane of the triangular system

blue components as the vertices of a *colour triangle*, as drawn in Figure 1.6b. Other colours can be specified by coordinates in the plane of the colour triangle. The location given by the coordinates corresponds to the amounts r, g and b making up the colour. The coordinates which specify the centre of the colour triangle represent the case when the three primary colours are mixed in equal amounts and indicates the colour *white*. Such representations are called *chromaticity diagrams*. These diagrams represent hue and saturation but not lightness, which must still be added as a third axis perpendicular to the chromaticity diagram if this information has to be displayed.

Colours on television screens are produced by additive coloration as the screen is composed of small dots of three different phosphors each of which shines with one of three primary colours when an electron beam falls onto it. Additive coloration is also used in the painting technique known as pointillism. In this method of painting, the image is built-up by placing small dots of colour onto the canvas, making sure that they do not overlap. When viewed from a distance of a few metres such pictures appear bright and dynamic.

The colour patterns on the wings of many butterflies and moths are produced in a similar way. The wings are tiled with a fine mosaic of scales, each of which reflects only one colour. The colour perceived by the eye is an additive colour arising from the numerous closely spaced scales. The range of colours which can be produced by rather a few basic pigments is remarkable. For example, some perceived purples arise from mixtures of black, white and red scales, while some greens arise from mixtures of yellow and black scales.

1.7. COLOUR DUE TO ABSORPTION: SUBTRACTIVE COLORATION

Absorption has been used for many centuries to produce colour. The colours perceived by the eye in which absorption and selective reflection or transmission are important are said to be due to subtractive colour mixing. For example, the colour of stained glass and the colours seen in ordinary colour filters are examples of colour production in this way. A red colour filter absorbs all colours except red and transmits only red light. Green leaves absorb red and blue light and reflect the green component of the incident white light. Colour due to absorption is caused by the fact that some of the incident wavelengths are more strongly attenuated than others.

Figure 1.7a shows the fraction of light transmitted as a function of wavelength for a commercial glass colour filter. The range of the visible spectrum is indicated above the transmittance curve. The filter absorbs red light strongly and transmits violet and blue-green light, as in Figure 1.7b. If the filter is held up to the light it will look blue-green. If the filter is crushed to a powder it will also appear to be blue-green, because the red light will still be absorbed but at least some of the remaining wavelengths transmitted by the grains will be reflected from the many surfaces in the powder back towards the viewer.

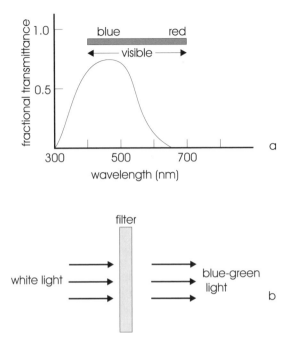

Figure 1.7 (a) The fractional transmittance of a commercial blue colour filter. About 3/4 of the blue light incident on the filter is transmitted but most red light is absorbed. (b). When the filter is viewed in transmitted white light it will appear blue-green

By analogy with additive coloration, one would expect to be able to combine three *subtractive primary colours* to produce the whole range of subtractive colours. These subtractive primary colours are *cyan*, which is red absorbing and transmits blue and green, *magenta*, which absorbs green and transmits blue and red and *yellow*, which is blue absorbing and which transmits green and red. If the three subtractive primaries are mixed we obtain *black*, as one primary will absorb red, one will absorb green and one absorb blue, thus removing the whole of the visible spectrum. The three dyes used in colour slide film and colour printers are the subtractive primary colours, cyan, magenta and yellow. This is called the CMYK colour model.

If the wavelength range of light absorbed is rather small, the colour remaining is called the *complementary* colour to that absorbed. Table 1.2 gives the colours and complementary colours of the visible spectrum. It is seen that the additive and subtractive primary colours are complementary colours.

Table 1.2 Complementary colours

Wavelength (nm)	Colour absorbed	Complementary colour
400–435	Violet	Yellow-green
435–480	Blue	Yellow
480–490	Green-blue	Orange
490–500	Blue-green	Red
500–560	Green	Magenta
560–580	Yellow-green	Violet
580–595	Yellow	Blue
595–605	Orange	Green-blue
605–700	Red	Cyan

1.8 THE INTERACTION OF LIGHT WITH A TRANSPARENT MATERIAL

Light can interact with a transparent material in several ways, shown schematically in Figure 1.8. The incident light can be *reflected* at any surface. The light passing through the material can be *scattered* or *absorbed*. If some of the absorbed light is re-emitted, usually at a lower energy, it is called *fluorescence*. The light which leaves the material is the *transmitted* light.

Ignoring fluorescence, these interactions can be expressed thus:

Incident intensity (I_o) = amount of light reflected (I_r)
+ amount scattered (I_s) + amount absorbed (I_a)
+ amount transmitted (I_t)

$$I_o = I_r + I_s + I_a + I_t$$

$$\text{or } 1 = R + S + A + T$$

where R is the *fraction* of light reflected, S the *fraction* of light scattered, A the *fraction* of light absorbed and T the *fraction* of light transmitted. In good quality optical materials the amount of light scattered and absorbed is small and it is often adequate to write:

$$I_o = I_r + I_t$$
$$1 = R + T$$

The appearance of a solid is often dominated by reflection. If the surface is smooth the reflection is said to be *specular* while if the surface is rough

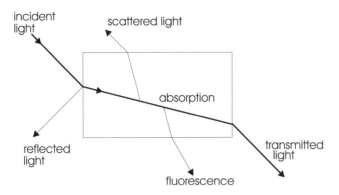

Figure 1.8 The interaction of light with a transparent material. The "bending" of the light ray on entering and leaving the material is called refraction and is discussed in Chapter 2. All of the processes labelled (reflection, absorption, scattering and fluorescence), can lead to colour production

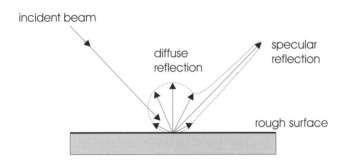

Figure 1.9 Reflection of light from a rough surface consists of two components, diffuse reflection and specular reflection. As the surface roughness increases the amount of diffuse reflection relative to the specular reflection increases. The ratio is an indication of surface gloss

it is *diffuse*, as depicted in Figure 1.9. The diffuse reflection component increases with surface roughness at the expense of the specular component so that a finely ground powder surface shows only diffuse reflection. The *gloss* of a surface is a measure of the relative amounts of diffuse reflectance to specular reflectance. Glossy surfaces have a large specular component.

Scattering can take place at crystallites or other inhomogeneities, called *scattering centres*, in the bulk. Although this is often undesirable, some glasses, such as *opal* glasses, are deliberately made with large numbers of scattering centres present. The resultant scattering renders the material

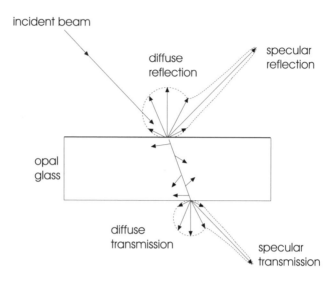

Figure 1.10 The passage of light through a translucent material containing many scattering centres, such as opal glass. Both the surface reflection and the transmitted light have diffuse and specular components

non-transparent. A material which allows light to pass through it but is not clear is said to be *translucent*. A light beam emerging from a translucent material will also consist of a diffuse and specular component, as illustrated in Figure 1.10.

In a discussion of absorption it is often useful to assume that the cause of absorption is due to the presence of *absorption centres*. These might be, for example, dye molecules, transition metal ions or small metal particles. When absorption centres are distributed uniformly throughout the bulk the amount of light absorbed in a plate is given by:

$$I = I_o \exp(-\alpha_a l) \tag{1.1}$$

where I is the intensity leaving the plate, I_o is the *incident intensity*, l is the thickness of the plate (m) and α_a is the *linear absorption coefficient* (m^{-1}). This expression is *Lambert's Law*. Note that the degree of absorption will vary significantly across the visible and many absorption centres show a pronounced maximum absorption at a particular wavelength λ_{max}.

The amount of absorption will undoubtedly be a function of the concentration of the absorbing centres throughout the bulk of the material. This is taken into account in the *Beer–Lambert law*:

$$\log (I / I_o) = -\varepsilon[J]\, l$$

where I is the intensity after passage through a length of sample l, I_o is the incident intensity and [J] is the *molar concentration* of absorption centres or absorbing species. The quantity ε is called the *molar absorption coefficient* (sometimes the *extinction coefficient*) and has units (dm^3 mol^{-1} m^{-1}) or (l mol^{-1} m^{-1}). The dimensionless product A = ε [J] l is called the *absorbance* (sometimes the *optical density*) and the ratio I/I$_o$ is the *transmittance*, T. Thus we can write:

$$\log T = -A$$

The Beer–Lambert law finds use in the measurement of concentrations. For example the clarity or otherwise of polluted air is often measured by comparing the intensity of light at a certain time (I) with the intensity on a fine day (I$_0$).

Reflection, scattering and absorption can all give rise to the world of colour around us, as illustrated in Plate 1.1. How this is brought about and how the colours can be put to use forms the substance of the following chapters.

1.9 ANSWERS TO INTRODUCTORY QUESTIONS

What is light?
For most purposes light can be treated as an electromagnetic wave with wavelengths varying between 400 nm and 700 nm, or as a stream of photons with an energy related to the wavelength as quoted by Planck's equation.

What is colour?
Colour is the physiological perception noted when light of a certain wavelength impinges upon the retina in the eye. For practical purposes, colour ranges from red at the long wavelength end of the spectrum (700 nm) through orange, yellow, green, blue, indigo to violet at the short wavelength end (400 nm).

Why do hot objects become red or white hot?
Hot objects emit a wide range of wavelengths of light. As the temperature of the body increases the wavelength with maximum intensity shortens.

At room temperature the peak wavelength is in the infrared, far beyond the visible spectrum, and none of the emissions can be seen. When the temperature of the body reaches about 1000 K the range of wavelengths emitted has a short wavelength tail which creeps into the red end of the spectrum and the body is said to be red hot. As the temperature increases the wavelength with the maximum intensity moves into the visible. The colour of the body changes from red to orange towards yellow. When the peak wavelength is close to the yellow region of the spectrum the body looks white to the eye and is said to be "white hot".

How do colour filters affect colour?
Colour filters "filter out" some colours. That is, filters absorb some wavelengths of light more than others. The colours seen when a beam of white light is transmitted through the filter are subtraction colours. (NB The colour of the filter looks quite different in reflected white light.)

1.10   FURTHER READING

Light described in terms of quantum electrodynamics is explained lucidly and non-mathematically in:

R. P. Feynman, QED: the Strange Theory of Light and Matter, Princeton University Press, Princeton (1985).

The fascinating history of the theories of light is given by:

G. N. Cantor, Optics after Newton, Manchester University Press, Manchester (1983).

A rigorous text which integrates ray optics, wave optics and quantum optics is:

B. E. E. Saleh and M. C. Teich, Fundamentals of Photonics, Wiley, New York (1991).

A comparison between the wave and particle explanation of the photo-electron effect and profound discussions of the relationship between particle and wave theories of atomic physics are given by:

D. Bohm, Quantum Theory, Prentice-Hall, Englewood Cliffs, NJ (1951).

Background to all of the material in this chapter is in:

K. Nassau, The Physics and Chemistry of Color, Wiley-Interscience, New York (1983), Chs 1 and 2.

The complexities of analysing colour and descriptions of the construction and use of chromaticity diagrams are detailed by:

J. P. Bouma, Physical Aspects of Colour, 2nd edition, eds W. de Groot, A. A. Kruithof and J. L. Ouweltjes, Philips Technical Library, Macmillan, London (1971).
C. S. Williams and O. A. Beckland, Optics, Wiley-Interscience, New York (1972), p. 368.

Colour vision is discussed by:

C. S. Williams and O. A. Beckland, Optics, Wiley-Interscience, New York (1972), p. 339.

1.11 PROBLEMS AND EXERCISES

1. Write a computer program to display the wave equation

$$y = a_0 \sin (2\pi/\lambda) (x + vt)$$

2. Extend the program in Q1 to add two waves

$$y_1 = a_0 \sin (2\pi/\lambda) (x_1)$$

$$y_2 = a_0 \sin (2\pi/\lambda) (x_2)$$

   varying x_2 between x_1 and λ, so as to reveal the consequences of constructive and destructive interference.

3. Plot the resultant intensity found in Q2, using $I = (y_1 + y_2)^2$.

4. Calculate the frequency and the energy of light photons with wavelengths of 425 nm, 575 nm and 630 nm. Describe the colour associated with each of each of these wavelengths.

5. Derive the equation

$$\rho(\nu)d\nu = 8\pi h\nu^3 \, d\nu / c^3 \, [\exp(h\nu/ kT) - 1]$$

   from that written in terms of λ given on page 9.

6. Show by dimensional analysis that the right-hand sides of the equations for $\rho(\lambda)d\lambda$ and $\rho(\nu)d\nu$ each have units of energy density (J m^{-3}).

7. Write a computer routine to plot the equation for black body radiation

$$\rho(\lambda)d\lambda = 8\pi hc \, d\lambda \, / \, \lambda^5[\exp(hc/\lambda kT) - 1]$$

in the form of $\rho(\lambda)d\lambda$ as a function of λ for several fixed values of T. Use this to determine how λ_{max} varies with T.

8. Spots of oil paint, red, green and blue, are arranged in a non-overlapping array. What would the colour be when viewed from a few metres away? The same three colours are mixed together on the artist's palette. What would the colour be?

9. The intensity of a light beam traversing a solution placed into a cell of 10 cm path length drops to 80.3 per cent of the incident intensity. Calculate the linear absorption coefficient of the solution.

10. How are the typical colours of tomatoes, oranges, black grapes and green leaves produced?

APPENDIX 1.1   ENERGY AND WAVELENGTH CONVERSIONS

A wide variety of energy units are used in the literature connected with light. A common non-standard unit of energy is the electron volt (eV), but spectroscopists more often use reciprocal centimetres (cm^{-1}). In wavelength designations, a common non-standard unit is the Ångström (Å).

To convert wavelength in nm to Å, multiply the values given by 10
To convert wavelength in Å to nm divide the values given by 10

To convert energy in J to eV, divide the values given by 1.60219×10^{-19}
To convert energy in eV to J, multiply the values given by 1.60219×10^{-19}

To convert energy in cm^{-1} to eV, multiply the value given by 1.2399×10^{-4}
To convert energy in cm^{-1} to J, multiply the value given by 1.9865×10^{-23}

To convert an energy in eV to the equivalent wavelength, use:
wavelength (nm) = 1239.9 / energy (eV)

Remember the standard SI prefixes indicate that:

to convert wavelength in nm to μm, divide the values given by 1000
to convert wavelength in μm to nm multiply the values given by 1000

The important constants for light are:

c, the speed of light in a vacuum	2.998×10^{8} m s^{-1}
h, Planck's constant	6.626×10^{-34} J s
k, Boltzmann's constant	1.380×10^{-23} J K^{-1}
e, charge on the electron	1.602×10^{-19} C
ε, permittivity of a vacuum	8.854×10^{-12} J^{-1} C^{2} m^{-1}

CHAPTER 2

COLOUR DUE TO REFRACTION AND DISPERSION

Why are images of objects in water displaced?
Why does lead added to glass make it sparkle?
Why do prisms produce a spectrum of colours?
How do rainbows form?

Many of the most beautiful colours arise simply by the interaction of light and transparent materials. Rainbows provide a familiar example. In the next few chapters the formation of these beautiful effects is explained. We start with a consideration of refraction and dispersion of light.

2.1 REFRACTION

When light enters a transparent medium it is *refracted* as shown in Figure 2.1. Refraction is the cause of the apparent bending of a ray of light when it enters water or glass. The magnitude of the effect is given by the *index of refraction* or *refractive index*, n, where:

$$n = \sin\theta_1 / \sin\theta_2$$

θ_1 being called the *angle of incidence* and θ_2 the *angle of refraction*. This equation is known as Snell's law. The *plane of incidence* is the plane containing the incident ray and the normal to the surface. The above equation is a special case of the more general relation:

$$\sin\theta_1 / \sin\theta_2 = n_2 / n_1$$

For light passing from a medium of refractive index n_1 to one of refractive index n_2.

The effect of refraction is familiar to anyone who has looked into a deep pool of water. In a swimming pool, for example, the bottom always seems closer to the surface than it really is. For the same reason, a stick will

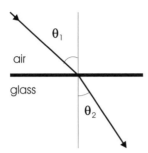

Figure 2.1. The refraction of a beam of light as it enters a block of a transparent medium such as glass with a high refractive index from a medium of low refractive index such as air. The light beam appears to bend at the interface by an amount given by Snell's law

appear bent towards the surface when dipped into water, as shown in Figure 2.2a. Kingfishers, birds which catch fish by diving into rivers, must allow for this effect and aim below the object that they apparently see in order to hit the target, as depicted in Figure 2.2b.

In effect, the refractive index is a manifestation of the fact that the light is slowed down on entering a transparent material. We shall see below that this is due to the interaction of the light with the electrons around the atoms which make up the solid. It is found that the refractive index, n, of a transparent substance is given by:

n = velocity of light in a vacuum (c) / velocity of light in the medium (v)

The frequency of the light does not alter when it enters a transparent medium and because of the relationship between the velocity of a light wave and its frequency:

$$\nu\lambda = \text{velocity}$$

it is possible to write:

$$n = c / v = \lambda_{vac} / \lambda_{subs}$$

where λ_{vac} is the wavelength of the light wave in a vacuum and λ_{subs} is the wavelength in the transparent substance. It is thus seen that light has a smaller wavelength in a transparent material than in vacuum, as illustrated in Figure 2.3.

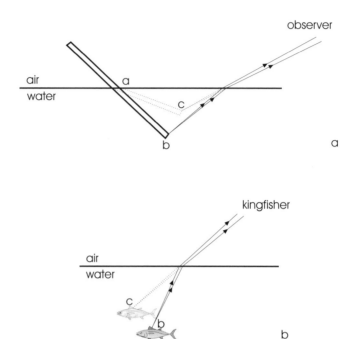

Figure 2.2 Because of refraction at the air–water surface submerged objects appear to be displaced when viewed by an observer above the surface. (a) A stick half immersed in water seems to bend upwards. Rays of light reflected from b appear to come from c. The stick appears bent at a. (b) An object such as a fish in a pool appears to be nearer the surface than it is. Light rays reflected from the fish at b, appear to to originate from c, causing the image of the fish to appear nearer to the surface than the actual fish

This can introduce confusion when the path of light rays through differ-ent materials has to be compared. To overcome this it is useful to define the *optical path* or *optical thickness* [d], and distinguish it from the *real* or *physical thickness* of a material, d. The relationship is given by:

$$[d] = nd$$

For several transparent materials traversed in sequence

$$[d] = n_1d_1 + n_2d_2 + n_3d_3 + \ldots$$

Referring to Figure 2.3, the physical thickness of the three slabs shown are all the same and equal to d while the optical thickness of the middle slab is twice that of the other two and is 2d.

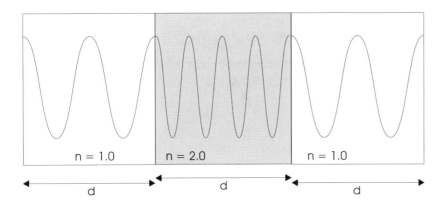

Figure 2.3 The effect of the refractive index on the wavelength of light. The wavelength in a medium of refractive index 2.0 is half that in a medium of refractive index 1.0

Note that in many important crystalline materials the index of refraction *varies with direction*. This interesting situation is taken further in Chapter 3. In the present chapter we will consider only those materials where this is not so and the index of refraction is the same regardless of the direction that the light takes.

2.2 TOTAL INTERNAL REFLECTION

When light passes from a higher refractive index material such as glass to one of lower refractive index such as air the refraction causes the emerging ray to bend towards the interface, as shown in Figure 2.4a. As the angle, θ, at which the ray approaches the surface increases the angle of the emerging ray becomes closer to the surface, as in Figure 2.4b. At the *critical angle*, θ_c, the emerging ray actually travels exactly along the surface, as in Figure 2.4c. If θ_c is exceeded then no light escapes and all behaves as if it were reflected from the under surface, shown in Figure 2.4d. This effect is called *total internal reflection*. In this case *no* light crosses the boundary and the light is trapped in the high refractive index medium.

The critical angle is given by the expression:

$$\sin \theta_c = n(\text{low}) / n(\text{high})$$

which follows from the general relation given above when θ_2 is equal to 90°.

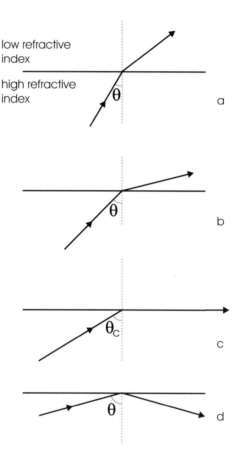

low refractive
index

high refractive
index

θ

a

θ

b

θ_c

c

θ

d

Figure 2.4 (a, b) When light passes from a medium of high refractive index (such as glass) to one of low refractive index (such as air) it will be refracted into a path which lies closer to the interface between the materials. (c) At a certain angle of incidence, the critical angle, θ_c, the ray will emerge along the surface itself. (d) For angles of incidence greater than the critical angle no light will emerge and total internal reflection will occur

This is the reason why a swimmer underwater will see the air surface as a bright "hole" in a surrounding dark continuum.

2.3 REFRACTIVE INDEX AND POLARISABILITY

To understand the relationship between refractive index and the atomic or molecular structure of a material it is necessary to remember that light can

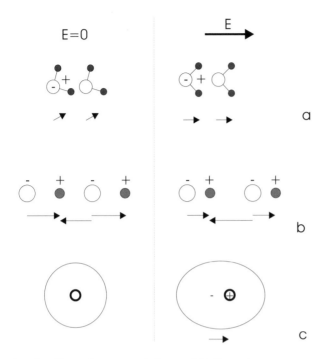

Figure 2.5 The effects of an external electric field, E, on the components of a solid. (a) Molecules with permanent dipoles will align in the field as much as the structure will allow. (b) Ions which are evenly spaced tend to be displaced in the field so as to create a net dipole moment. (c) The uniform electron cloud around an atom or an ion tends to distort so as to produce a dipole. The displacements have all been grossly exaggerated for clarity and the dipoles are indicated as arrows with the positive end represented by an arrowhead

be treated as a varying electric field. If a static electric field is applied to an insulating material, the internal components which carry a charge will try to line up with the field and the material is said to become *polarised*. The most important of the charged internal components are: (i) the permanent molecular dipoles present; (ii) the positive and negative ions present; and (iii) the electrons present. In a static electric field existing molecular dipoles will reorient themselves in the field as much as the surroundings will allow, illustrated in Figure 2.5a. In the case of ionic solids the overall dipole moment is zero in the absence of an electric field. In a static electric field the ions move slightly so as to produce an net dipole moment, illustrated in Figure 2.5b. The lightest component, the negatively charged electron cloud surrounding the atomic nucleus is easily deformed by an external field to create a dipole, as sketched in Figure 2.5c.

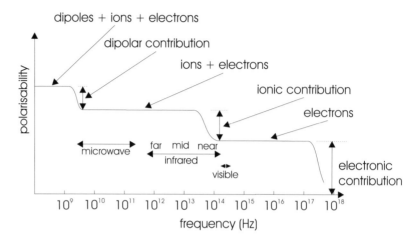

Figure 2.6 A schematic illustration of the contribution of permanent dipoles, ions and electrons to the total polarisability of a material as the frequency of the applied field is increased. The contribution due to permanent molecular dipoles is lost when the field frequency reaches the microwave region and the contribution of the ions is lost at near infrared frequencies. Only the effect of electronic polarisability occurs at optical frequencies

If the electric field is not static, but consists of an alternating field, the dipoles, ions and electrons will try to follow the changes in the field direction and move to and fro. This effect is utilised in microwave ovens, which bombard the contents with radiation at frequencies of about 10^9 Hz. As the applied electric field changes direction, the dipoles, especially those associated with water molecules, reorient to and fro. This continuous motion heats the food in the oven. However, motion is restricted for molecular dipoles and when the frequency of the applied electric field becomes much higher than that of microwaves (of the order of 10^{10} Hz) any contribution of the dipoles is lost as the electric field is now changing too rapidly for them to keep up. The magnitude of the polarisability thus falls to a lower plateau, as shown in Figure 2.6. When the frequency of the field reaches that of near infrared radiation (approximately 10^{14} Hz) even the lightest ions can no longer move to and fro quickly enough and their contribution to the polarisability is now lost. The magnitude of the polarisability then falls to a lower plateau, see Figure 2.6. The electrons, however, can follow the oscillations of a varying electrical field even at visible and ultraviolet frequencies, and it is these which are most important in colour production. This response of the electrons to an applied alternating electric field is called the *electronic polarisability*.

The bulk polarisation, P, is given by the equation:

$$P = \varepsilon_0 \chi E$$

where ε_0 is the permittivity of free space, χ is the dielectric susceptibility and E the applied electric field. The refractive index of a material, n, is actually a reflection of its electronic polarisability. The relationship between the two is given by the equations:

$$n \sim \sqrt{\varepsilon_r}$$

where

$$\varepsilon_r = \chi + 1$$

and ε_r is the *relative permittivity* of the material, called the *dielectric constant* in older literature.

In general, strongly bound electrons, trapped at atomic nuclei or in strong chemical bonds have a low polarisability and this leads to a low refractive index. Loosely bound electrons such as outer electrons on large atoms, or lone pair electrons, are highly polarisable and so will yield materials with a larger refractive index.

This effect is well known. For example, lead oxide, PbO, contains large Pb^{2+} ions with a highly polarisable lone pair on each Pb^{2+} ion. When lead oxide is added to ordinary glass the highly polarisable Pb^{2+} ions (which occupy positions between the chains of SiO_4 tetrahedra making up the structure), have the effect of considerably increasing the refractive index of the glass. Flint glass, which contains significant amounts of lead oxide, is therefore prized and used as "cut glass" and lead "crystal" because the higher refractive index gives a more attractive appearance to the articles. Table 2.1 shows the effect of added PbO on the refractive index of three different flint glasses.

2.4 REFRACTIVE INDEX AND DENSITY

Electronic polarisability is not the only factor which contributes to refractive index. A little thought will show that the number of atoms, ions or molecules present is also important. Gases, because of their low densities, have refractive indices close to 1. However, although small, the variation of

Table 2.1 Some refractive indices

Substance	Refractive index[a]	Substance	Refractive index[a]
(Vacuum)	1.0 (definition)	Dry air, 1 atm 15°C	1.00027
Water	1.3324	Na_3AlF_6 (cryolite)	1.338[b]
MgF_2 (sellaite)	1.382[b]	Fused silica (SiO_2)	1.4601
KCl (sylvite)	1.490	Crown glass	1.522
Extra light flint glass[c]	1.543	NaCl (halite)	1.544
Flint glass[c]	1.607	MgO (periclase)	1.735
Dense flint glass[c]	1.746	Al_2O_3 (corundum)	1.765[b]
ZrO_2 (baddeleyite)	2.160[b]	C (diamond)	2.418
$CaTiO_3$ (perovskite)	2.740	TiO_2 (rutile)	2.755[b]

[a] A value appropriate to the yellow light emitted by sodium atoms (the sodium D-lines, see Chapter 7), with an average wavelength 589.3 nm, is given.
[b] The refractive index varies with direction; the average value is given.
[c] The flint glasses contain significant amounts of lead oxide, PbO, as follows: extra light flint, 24 mass% PbO; flint, 44 mass% PbO; dense flint, 62 mass% PbO.

the density of air as a function of temperature is the source of mirages and related visual effects. In general, a temperature gradient in the air changes the refractive index of the air and sets up an "air lens". Because the lens is imperfect the images reaching the eye are imprecise and the human imagination has constructed a variety of fanciful explanations for the apparitions observed. These include not only the well-known water pools in the desert, but Atlantis myths and supposed walking on water. A medium with a varying refractive index is called a *graded index* (GRIN) material. We will meet with graded index glass in the context of optical fibres in Chapter 11.

Densely packed arrays of atoms in liquids and solids have a higher refractive index than gases. Although the refractive indices of most simple compounds are known it is sometimes useful to estimate the refractive index of more complex or hypothetical materials. One of the most successful ways of doing this is via the *Gladstone–Dale formula*, which combines density and, indirectly, polarisability terms. It is especially useful for complex oxides, for which the Gladstone–Dale formula can be written:

$$n = 1 + \rho(p_1k_1 + p_2k_2 + p_3k_3 \ldots)$$

$$\text{or} \quad n = 1 + \rho\sum p_i k_i$$

where ρ is the density of the complex oxide and the terms p_i and k_i are defined below. The assumption underlying the formula is that the refractive index of a complex oxide is made up by adding together the contributions from a collection of simple oxides, oxide 1, oxide 2 and so on, for which optical data are known. The polarisability is taken into account by allocating to each of the simple oxide components a factor k called the *refractive coefficient*, an empirically determined constant. The amount of each oxide is taken into account by multiplying the refractive coefficient by its weight fraction in the compound, p. A number of values of k for use in the Gladstone–Dale formula are given in Table 2.2.

The rule works well and usually gives answers within about 5 per cent. Note, however, that the value obtained is an average refractive index. Many oxides have refractive indices which vary according to crystallographic direction. The Gladstone–Dale relationship ignores this feature.

The equation can also be used to determine a value of either density or average refractive index for unknown polymorphs of simple oxides. For example, the Gladstone–Dale equation for the polymorphs of SiO_2 is:

$$n = 1 + 0.21\rho$$

and for the polymorphs of TiO_2 is:

$$n = 1 + 0.40\rho$$

Provided the density of each polymorph is known its average refractive index can be found and vice versa.

2.5 DISPERSION AND COLOUR PRODUCED BY DISPERSION

The refractive index of a transparent solid varies with wavelength. This is called *dispersion*. In general the index of refraction increases as the wavelength decreases so that the refractive index of red light in a material is less than that of violet light. The dispersion can be formally defined as $dn/d\lambda$, which is the slope of the refractive index, n, versus wavelength, λ, curve. Although the dispersion of many materials is rather small, it is important to include it when calculating the optical properties of lenses and similar high quality optical components. The refractive indices of fused silica glass and corundum, Al_2O_3, both important optical materials, are given as a function of wavelength in Figure 2.7.

Table 2.2 Refractive coefficients for some oxides

Oxide	k	Oxide	k	Oxide	k	Oxide	k	Oxide	k	Oxide	k	Oxide	k	Oxide	k
H_2O	0.34	BeO	0.24												
Li_2O	0.31	MgO	0.20			TiO_2	0.40			B_2O_3	0.22	SiO_2	0.21	P_2O_5	0.19
Na_2O	0.18	CaO	0.23	Y_2O_3	0.14	ZrO_2	0.20	Nb_2O_5	0.30	Al_2O_3	0.20				
K_2O	0.19	SrO	0.14	La_2O_3	0.15							SnO_2	0.15	Bi_2O_3	0.16
		BaO	0.13									PbO	0.15		

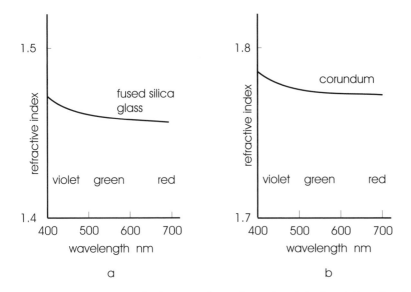

Figure 2.7 The variation of refractive index with wavelength for fused silica glass (a) and corundum, Al_2O_3 (b). In the case of corundum the refractive index depends upon direction and average values plotted

For many transparent materials a good representation of the variation of refractive index with wavelength in the visible region is given by Cauchy's equation,

$$n = A + B/\lambda^2 + C/\lambda^4$$

where A, B and C are empirically determined parameters. For lens design Cauchy's equation is not sufficiently precise and a more accurate formula, which gives the refractive index of glasses in the wavelength range 365–2300 nm to high degree of fidelity is:

$$n = \{1 + [B_1\lambda^2 / (\lambda^2 - C_1) + B_2\lambda^2 / (\lambda^2 - C_2) + B_3\lambda^2 / (\lambda^2 - C_3)]\}^{1/2}$$

where the wavelength λ is in μm, and B_1 to C_3 are constants appropriate to the glass. (Some typical values for the constants will be found in problem 7 at the end of this chapter.)

A widely used measure of the dispersion of a material, is the *Abbé v-value*, written v_D, and given by:

$$v_D = (n_D - 1) / (n_F - n_C)$$

where n_d is the refractive index of the material at a (yellow) wavelength, usually taken as 587.56 nm (the helium D-line, see Chapter 7), n_F is the refractive index of the material at a (blue) wavelength of 486.1 nm and n_C is the refractive index of the material at a (red) wavelength of 656.3 nm. The reciprocal of the Abbé v-value is often called the *dispersive power*.

The formation of a spectrum by the dispersion of white light with a glass prism, surely one of the most exciting of experimental results, was first interpreted as such by Newton. In fact it is the differing values of refractive index for different wavelengths, i.e. the dispersion by the glass in the prism, which produces the effect, as shown in Figure 2.8a. Snell's law tells us that for a given angle of incidence, θ_i, $\sin \theta_r$ is inversely proportional to the refractive index, n, so that as n increases θ_r decreases and the ray deviates more. Red light then tends to be the least deviated and violet light the most. The higher the dispersion, the wider will be the spectrum.

Exactly the same effect is found in simple lenses. The edge of the lens is approximately prism shaped and dispersion causes the image to become coloured at the periphery of the field of view, as shown in Figure 2.8b. This effect is known as *chromatic aberration*. If the lens is considered to be a thin prism, the angle of deviation of the rays, δ, will be given by:

$$\delta = (n_\lambda - 1)\alpha$$

where n_λ is the refractive index appropriate to the colour, and α is the small angle at the top of the prism, sketched in Figure 2.8c. Chromatic aberration is avoided in expensive lenses by using combinations of glasses chosen so as to eliminate the effects of dispersion in each component. Such compound lenses are called *achromats*.

Dispersion is responsible for the flashes of colour, known as *fire*, that are such an important feature of diamonds. The production of the colour is shown in Figure 2.9. In diamond it is the combination of very high refractive index and high dispersion that causes the stones to be prized. The stones are "cut" (actually cleaved) so as to produce many facets, each of which can act as a tiny prism, thus greatly enhancing the display of fire as the gem moves.

2.6 RAINBOWS AND HALOS

The rainbow is one of the most beautiful examples of colour produced by refraction. Most frequently seen, when the observer's back is to the sun, is a

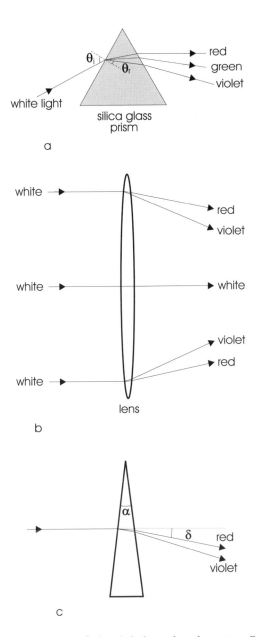

Figure 2.8 (a) The refraction of white light by a silica glass prism. For silica glass, the refractive index for violet light is greater than for red light (Figure 2.7), which disperses the light to form a spectrum. (b) For the same reason the edge of a simple lens acts as a prism and so causes chromatic aberration. (c) The deviation of light by a thin prism provides a useful model for the dispersion from a thin lens. Each colour will be deviated by a different amount, given by $\delta = (n_\lambda - 1)\alpha$

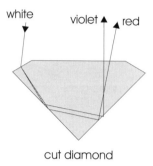

Figure 2.9 The combination of high refractive index and high dispersion in cut diamonds gives these gemstones the ability to produce spectral colours, known as fire

single bright arc called the *primary rainbow*. The colour violet is always innermost and the colours proceed through indigo, blue, green, yellow, orange and red on the outside of the arc. The acute angle between the observed direction of the bow and that of the sunlight producing it is about 42°, as depicted in Figure 2.10a. The locus of this angle generates the arc seen and the observer appears to be at the apex of a cone with a semi vertical angle of 42°. A careful examination of the sky near a rainbow will often, but not always, show many other features including a fainter *secondary rainbow* at an angle of about 50° and various *supernumerary arcs* inside the primary bow.

Although a complete description of all of these features is complex, the explanation of the primary rainbow is relatively simple. It is produced by a single reflection from inside a raindrop, as shown in Figure 2.10b. The violet light emerges from each raindrop at an angle of 41° and the red light at an angle of 43°, with the intermediate colours lying between these limits. As can be appreciated from Figure 2.10b these rays diverge. This means that each of the colours that enter the eye from a rainbow originates in a separate arc of water drops. For the same reason, no two observers ever see exactly the same rainbow. Each person sees only a unique part of the raindrop curtain that subtends the correct angles with respect to the observing eye.

Other rainbows are produced by more internal reflections within the water drops. One can frequently see a fainter secondary bow, produced by two internal reflections, lying outside the primary bow. Three internal reflections produce a ternary bow, and so on, but these are not usually observed in nature. However, up to a dozen rainbows can be seen in a single drop of water indoors if carefully observed.

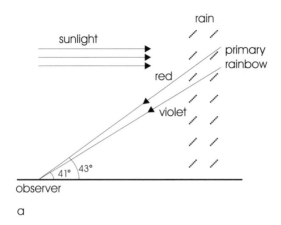

a

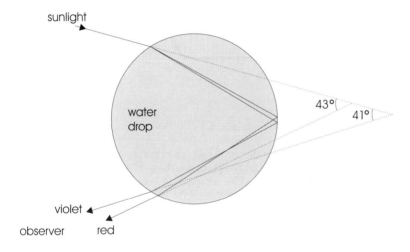

b

Figure 2.10 The refraction of white light by raindrops is responsible for the formation of rainbows. To observe a rainbow, sunlight must come from behind the observer. (a) In the main or primary bow the violet light appears to come from a cone of angle 41° and the red light from a cone of larger angle, 43°. (b) The primary bow is produced by a single reflection within each raindrop combined with dispersion of the light due to the variation of refractive index with wavelength

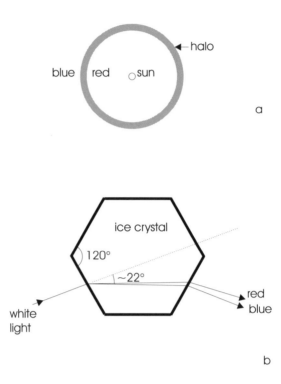

Figure 2.11 (a) A halo is a diffuse ring of light, often red on the inner side and blue on the outer side, seen around a bright object such as the sun. The total angular width of a halo is 44°. (b) The halo is formed by refraction of light through a random array of hexagonal ice crystals. The average deviation of the light in each crystal is 22°, with the deviation of the red ray about 1.5° less than that of the blue. (The angles are exaggerated for clarity)

Halos are not as spectacular as rainbows. They are often observed as a rather pale white or coloured ring around the sun or moon. Once again refraction and dispersion are responsible for the colour, but in this instance the refraction occurs within randomly oriented hexagonal ice crystals in the upper atmosphere, as depicted in Figure 2.11a. The commonest halo, the 22° halo, is a ring subtending a semi-angle of 22° to the observer's eye. Dispersion of the blue wavelengths is more than that of the reds and so the halo is red internally and blue-violet externally, as indicated in Figure 2.11b. Plate 2.1 shows a halo-like arc of colour due to refraction and dispersion in ice crystals in stratospheric clouds.

2.7 ANSWERS TO INTRODUCTORY QUESTIONS

Why are images of objects in water displaced?

The images of objects in water appear to be displaced because the rays of light reflected from the object into the eye of the observer travel at a different speed in water than in air. This makes objects seem to be nearer the surface than they really are. This effect is quantified by the index of refraction of water, which is equal to the ratio of the velocity of light in a vacuum (or, approximately, air) to that in water.

Why does lead added to glass make it sparkle?

Lead in the form of Pb^{2+} is added to glass to increase its refractive index. Pb^{2+} ions are called "lone pair" ions. They have a pair of electrons in an orbital which is not used in forming chemical bonds (a non-bonding orbital) and which projects out from the nucleus of the atom. These lone pair electrons are very exposed to applied electric fields and so are very polarisable. It is this high polarisability which results in a higher refractive index and a higher "sparkle" for the glass.

Why do prisms produce a spectrum of colours?

The refractive index of a substance is a function of the wavelength of the light. In general the refractive index of a transparent material is greater for shorter wavelength light (towards violet) than for longer wavelengths (towards red). A prism will cause red light to deviate least and violet light to deviate most, with intermediate wavelengths falling between these limits.

How do rainbows form?

Rainbows form when sunlight which comes from behind the observer enters a raindrop and is totally internally reflected once to reach the eye. The white light is spread out into a spectrum by the same effect as discussed in the previous question.

2.8 FURTHER READING

The physics of the rainbow, halos and other dispersion colours found in nature is given by:

D. K. Lynch and W. Livingston, Color and Light in Nature, Cambridge University Press, Cambridge (1995), Ch. 4.

H. Moysés Nussenzveig, Scientific American, **236**, April 1977, p. 116.

How to observe multiple rainbows in a single drop of water is explained by:

J. Walker, Scientific American, **237**, July 1977, p. 138.

The relationship between refractive index, polarisability and the Gladstone–Dale equation is given in:

F. D. Bloss, Crystallography and Crystal Chemistry, Holt Rinehart and Winston, New York (1971), Ch. 11.
R. E. Newnham, Structure–Property Relations, Springer, Berlin (1975), Ch. 5.

Mirages are described by:

A. B. Fraser and W. B. Mach, Scientific American, **234**, January 1976, p. 102.

2.9 Problems and Exercises

1. Determine the critical angle for a heavy-metal fluoride glass with a refractive index of 2.16 which is being evaluated for optical fibre purposes. What would the critical angle be if the fibres were coated in plastic with a refractive index of 1.32?

2. Estimate the refractive index of the garnet $Mg_3Al_2Si_3O_{12}$ which has a density of 3.56 g cm^{-3}.

3. Estimate the refractive index of beryl, $Be_3Al_2Si_6O_{18}$, with a density of 2.64 g cm^{-3}.

4. Estimate the density of ackermanite, $Ca_2MgSi_2O_7$, with a refractive index of 1.636.

5. Estimate the refractive coefficient of Al_2O_3 in andalusite, Al_2SiO_5, with a density of 3.15 g cm^{-3} and a refractive index of 1.639.

6. The average refractive indices of the two commonest polymorphs of TiO_2 are, rutile, 2.755 and anatase, 2.525. The density of rutile is 4.245 g cm^{-3} and of anatase is 3.923 g cm^{-3}. Estimate the refractive index of the rarest form of TiO_2, brookite, given that its density is 4.119 g cm^{-3}.

7. An optical glass has constants $B_1 = 1.040$, $B_2 = 0.2318$, $B_3 = 1.01047$, $C_1 = 0.006$, $C_2 = 0.020$, $C_3 = 103.56$. Calculate the Abbé v-value using the formulae in Section 2.5.

8. Calculate the width of the spectrum generated by the edge of a simple lens with refractive indices 1.513 (red, 700 nm) and 1.530 (violet, 400 nm) at a distance of 50 cm.

9. It is not uncommon to see two rainbows (a double rainbow), a faint one outside a strong one. How do you think that this arises? Apart from its intensity, how does the secondary bow differ from the single (primary) bow?

CHAPTER 3
CRYSTALS AND LIGHT

Why do some crystals produce double images?
Why do the colours of some crystals depend upon the viewing direction?
How can some prisms produce either one spectrum or two?
How can infrared radiation of wavelength 1.08 μm be changed into green light?
What are liquid crystals?

The interaction of crystals and light has long produced fascinating and puzzling experimental results. Many of these are bound up with the polarisation of light and the two topics cannot easily be separated. The polarisation of light does not produce colour by itself. However, mineralogists have long made use of the colours displayed by crystals in the polarising microscope for identification purposes. Colour is usually induced in polarised light by the phenomena of dispersion, absorption and interference. However, colour can also be produced by crystals using frequency doubling and a discussion of this and related phenomena are included. Finally, a description of liquid crystals and optical activity, which are relevant to colour displays, ends the chapter.

3.1 POLARISATION OF LIGHT

Light, as was mentioned in Chapter 1, can be regarded as a wave of wavelength λ with electrical and magnetic components lying at right angles to one another, each described by a vector. For the present only the electric field vector is important. This vector is always perpendicular to the line of propagation of the light but can adopt any angle otherwise, similar to the positions available to a hand on a clock. For ordinary light, such as that from the sun, the orientation of the electric field changes in a random fashion every 10^{-8} s or so, as if the seconds hand of a clock jumped

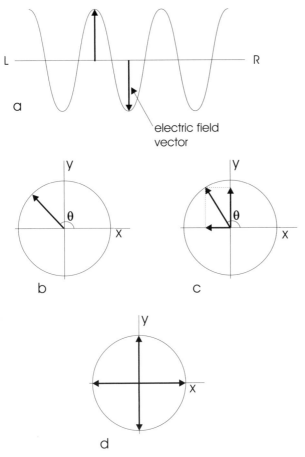

Figure 3.1 (a) A light wave moving from left to right with the electric field vector (bold arrows) permanently in the plane of the paper is said to be linearly polarised. (b) When viewed along the ray the electric field vector in a linearly polarised beam can take any orientation θ, with the direction of propagation. (c) The electric field vector at any instant can be resolved into two perpendicular components along x and y. (d) Over an interval, normal light can be resolved into components along Ix and Iy, with equal amplitudes but lacking coherence of phase

unpredictably from position to position without rotating in a steady manner. The position of the electric field defines the *polarisation* of the light wave. Ordinary light is said to be *unpolarised*.

For many purposes the polarisation of a light wave is important. Light is said to be *linearly* or *plane polarised* when the electric field which describes the light wave is forced to vibrate in a single plane. This is analogous to the seconds hand on a clock being stuck permanently in one position. When considering a light wave such as that shown in Figure 3.1a, propagating from left to right, the electric field can be considered to lie in any plane perpendicular to L-R. In the figure the vector is drawn in the plane of the page, but it could lie at any angle θ to the propagation direction, as indicated in Figure 3.1b. The light is linearly polarised if θ remains constant. (In this book the shorthand term "polarised light" will be taken to mean linearly polarised light.) It is often convenient to represent polarised light by resolving the electric field vector into two components along mutually perpendicular axes, x and y, as in Figure 3.1c. In the case of ordinary unpolarised light, because the direction of the electric field vector changes frequently, the time averaged vector can take any orientation. Thus in Figure 3.1b the angle θ would change at random in this way. In this case, the net result, taken over an interval of time, is equivalent to four vectors of equal lengths, at right angles to one another, but lacking phase coherence, as in Figure 3.1d.

3.2 POLARISATION BY REFLECTION

When light is incident upon the surface of a transparent dielectric such as glass, part will be reflected and part refracted, as drawn in Figure 3.2. This seemingly simple and familiar occurrence can produce polarised light. Taking the plane which contains the incident ray, the normal to the surface and the reflected ray as the plane of incidence, the component of the light wave polarised in the plane of incidence, called the p-wave, is reflected to a different extent than the component polarised perpendicular to the plane of incidence, called the s-wave. The difference is dependent upon the angle at which the light falls onto the surface (the angle of incidence, θ_1 in Figure 3.2). For many angles of incidence the reflection of the p-wave is somewhat suppressed relative to that of the s-wave. This causes the reflected light to be noticeably polarised. When this occurs the refracted part of the incident light will also be partly polarised.

With reference to Figure 3.2, the *reflectivity* or *reflectance* of the surface is given by:

$$R_s = [\sin(\theta_1 - \theta_3) / \sin(\theta_1 + \theta_3)]^2$$
$$R_p = [\tan(\theta_1 - \theta_3) / \tan(\theta_1 + \theta_3)]^2$$

for the s-wave and p-wave respectively. (Remember that θ_1 is equal to θ_2 for reflection.) The reflectivity of these two components at a glass surface is

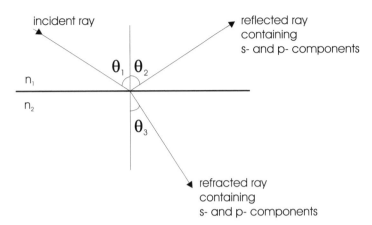

incident ray

reflected ray
containing
s- and p- components

θ_1 θ_2

n_1

n_2

θ_3

refracted ray
containing
s- and p- components

Figure 3.2   The geometry of reflection that leads to the production of polarised light. Note that $\theta_1 = \theta_2$ and both the reflected and refracted rays generally contain s-wave and p-wave components

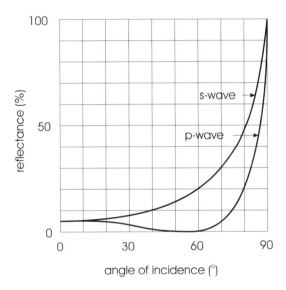

100

s-wave

reflectance (%)

50

p-wave

0

0   30   60   90

angle of incidence (°)

Figure 3.3   Reflection at a glass (n = 1.52) surface in air showing the p-wave and s-wave components. The reflectivity of the p-wave component is zero at Brewster's angle, ~57° for a glass/air interface

drawn in Figure 3.3. This shows that when light passing through air falls perpendicularly onto the surface of the material, with refractive index n, the angle of incidence, θ_1 is zero and the reflectivity of the p-wave equals that of the s-wave. The reflectivities of the s-wave and p-wave remain almost the same up to an angle of incidence, θ_1, of about 20°. Thereafter, the reflectivity of the s-wave smoothly increases to 100% at grazing incidence ($\theta_1 = 90°$). The reflectivity of the p-wave diverges from that of the s-wave and decreases as the angle of incidence increases until, at a particular angle, it becomes zero. At this point the reflected beam is polarised to its maximum extent. This optimal angle of incidence is given by *Brewster's law*:

$$\tan\theta_1 = \tan\theta_2 = n_2 / n_1$$

where the angles are given in Figure 3.2, n_1 is the refractive index of the initial medium that the light ray traverses and n_2 is the refractive index of the medium causing reflection. For glass with a refractive index of 1.52 in air ($n_1 = 1$ and $n_2 = 1.52$), Brewster's angle will be 56.7°. This situation is illustrated in Figure 3.4, in which polarisation perpendicular to the plane of the page is represented by filled circles along the ray and polarisation normal to this direction as double headed arrows. When both polarisation modes are present the symbols are superimposed. The relative amounts of the two modes is represented schematically by the lengths of the double headed arrows in the circles. As the angle of incidence increases past Brewster's angle the reflectivity of the surface for the p-wave will increase smoothly to 100 per cent at grazing incidence ($\theta_1 = 90°$). Thus at both perpendicular and grazing incidence the s-wave and p-wave behave identically and all of the light is reflected.

When total internal reflection is considered the reflectivities of the s-wave and p-wave components of the light will also be angle dependent in the same way. The reflectivity of the p-wave will become zero at the "internal" Brewster's angle of (90−56.7)° for glass, i.e. 33.3°. This has important consequences for the long distance performance of optical fibres.

3.3 POLARISATION USING POLARS

Polars are devices which transmit light vibrating (mainly) in a single plane. This plane is referred to as the *vibration direction* or the *allowed direction*.

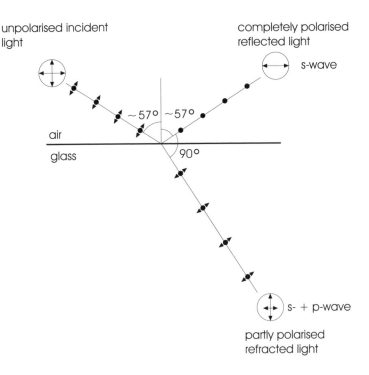

unpolarised incident
light

completely polarised
reflected light

s-wave

air

glass

90°

s- + p-wave

partly polarised
refracted light

Figure 3.4 Unpolarised light on reflection from a glass plate at Brewster's angle (~57°) will produce a completely linearly polarised s-wave and a partially polarised refracted wave. The two components of polarisation (see Figure 3.1d) are shown as double headed arrows or filled circles on the light rays. The relative amounts of each component as if viewed along the ray are shown by the double headed arrows in the open circles

Light can thus be made into a linearly polarised wave by passing it through a polar, as illustrated in Figure 3.5. Polars can be made from prisms of certain crystals or from sheets of certain organic molecules. *Polaroid*, familiar because of commercial use in Polaroid sunglasses, is such a material.

If two polars are arranged in tandem so that light passes through both, as shown in Figure 3.5, the first polar is called the *polariser* and the second the *analyser*. The arrangement is found in petrographic optical microscopes. In these instruments the interference colours produced when a thin rock sample is inserted between the two polars can be used to identify the mineral.

The light intensity transmitted by a pair of polars was first investigated some 200 years ago by Malus. When the vibration directions of polariser

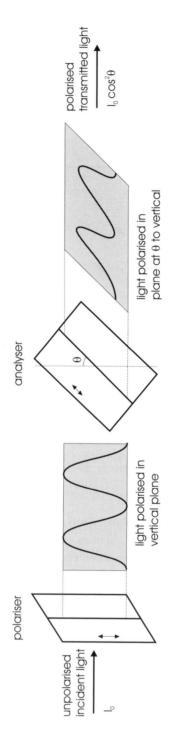

Figure 3.5 Normal light of intensity I_0 transmitted by a polar (at left) will emerge linearly polarised parallel to the vibration direction of the polar, marked as a double headed arrow and a heavy line. When two polars are arranged in sequence, the first polar is called the polariser (at left) and the second the analyser (at right). In such a case a beam transmitted by both polariser and analyser will have a linear polarisation parallel to the vibration direction of the analyser. The intensity will be given by $I_0 \cos^2 \theta$. If the vibration directions of the polariser and analyser are perpendicular to each other no light will be transmitted by the pair

and analyser are parallel the light transmitted will consist of a linearly polarised wave with an intensity equal to that of the incident radiation. If the analyser is now rotated with respect to the polariser the intensity transmitted will diminish according to the *Law of Malus*:

$$I = I_o\cos^2\theta$$

where I_o is the incident intensity and θ is the angle between the vibration directions of the polariser and analyser. The emergent wave will be linearly polarised in a plane corresponding to the vibration direction of the analyser, as sketched in Figure 3.5. No light will be transmitted when the vibration directions of the two polars are perpendicular to each other. In this orientation the polars are said to be *crossed*.

This result is put to practical use in "Polaroid" sunglasses, where the molecules in the film are arranged so as to endow it with a vertical vibration direction. As we have seen in the previous section, reflected light contains a considerable proportion of light polarised parallel to the reflecting surface, the s-wave component. The Polaroid sunglasses eliminate this horizontal component and so reduce glare considerably. Similar devices are used in photography to reduce the glare from surfaces of water or polarisation due to scattering (see Chapter 5).

3.4 CRYSTAL SYMMETRY

Gases, most liquids and some solids such as glasses are *isotropic* with respect to their refractive index. That is to say, the refractive index is the same in all directions. This is not generally true for crystalline materials. The optical behaviour is found to depend upon the symmetry of the crystal. Note that here it is the internal symmetry which is important and not the external shape, called the *morphology* or *habit*. Crystallographically, symmetry is defined in terms of *symmetry operators*, which apply reflections, rotations and so on, to the atomic and molecular components making up the crystal. From among the various symmetry operators, the presence of a centre of symmetry is of considerable significance from the point of view of optical properties. A centre of symmetry, at 0,0,0, transforms any point (x,y,z) to (−x,−y,−z). Both crystals and molecules which do not have a centre of symmetry are termed *non-centrosymmetric*.

Table 3.1 The crystal systems

Crystal system	Unit cell characteristics	Example
Cubic (isometric)	$a = b = c$, $\alpha = \beta = \gamma = 90°$	Halite, NaCl
Tetragonal	$a = b \neq c$, $\alpha = \beta = \gamma = 90°$	Rutile, TiO_2
Hexagonal	$a = b \neq c$, $\alpha = \beta = 90°$, $\gamma = 120°$	Zincite, ZnO
Trigonal[a]	$a = b = c$, $\alpha = \beta = \gamma \neq 90°$ or	Calcite, $CaCO_3$
	$a' = b' \neq c'$, $\alpha = \beta = 90°$, $\gamma = 120°$	Dolomite, $CaMg(CO_3)_2$
Orthorhombic	$a \neq b \neq c$, $\alpha = \beta = \gamma = 90°$	Stibnite, Sb_2S_3
Monoclinic	$a \neq b \neq c$, $\alpha = \gamma = 90°$, $\beta \neq 90°$	Tungsten trioxide, WO_3
Triclinic	$a \neq b \neq c$, $\alpha \neq \beta \neq \gamma \neq 90°$	Copper sulphate, $CuSO_4.5H_2O$

[a] Trigonal crystals are most often described in terms of an alternative hexagonal unit cell given in the second line of this category.

For our purposes the collection of symmetry operators to be found in a crystal is best described in terms of the *unit cell* of the material. The unit cell of a crystal is the smallest convenient volume of crystal which displays the symmetry of the crystal and, if extended in three directions (like building up a cube or pyramid from bricks), will produce the macroscopic crystal. It is characterised by three axes, labelled a, b and c, and the angles between them, α, β and γ, where α lies between b and c, β lies between a and c and γ between a and b. For historical reasons, the overall symmetry is referred to as the crystal *system* and given a name, orthorhombic for example, which is related to the shape of the crystallographic unit cell. For example, a crystal possessing orthorhombic symmetry will be shaped like a brick, with three unequal axes all at 90° to each other. There are seven possible crystal systems, given in Table 3.1.

Cubic crystals like common salt (halite or rock salt) have the same refractive index in all directions and behave in the same way as a glass with respect to light. In all of the other classes this is not so. In tetragonal, trigonal and hexagonal crystals the refractive indices along the a- and b-axes are the same and different from the refractive index along the c-axis. In ortho-rhombic, monoclinic and triclinic crystals there are three refractive indices along mutually perpendicular axes. This is unsurprising. The refractive index depends upon the density of atoms in a crystal. In cubic crystals the atom density averages to be the same in all directions while in crystals of lower symmetry some directions contain more atoms than others. For example, in the tetragonal rutile structure of TiO_2 chains of TiO_6 octahedra run along the c-axis. This structural feature results in the atoms in the crystal being much less densely packed along the a- and b- axes than along the c-axis chains. The refractive index along a- and b- is 2.609 while along c- it is 2.900.

3.5 DOUBLE REFRACTION BY CRYSTALS

Although a variation in refractive index with direction may not be surprising, the way in which crystals with structures other than cubic interact with light is certainly so. We will illustrate this by reference to the mineral calcite.

Calcite is a mineral form of calcium carbonate, $CaCO_3$. The unit cell is trigonal with a = 0.641 nm and α = 101.9°, but it is more convenient to refer to a hexagonal unit cell in which a = 0.499 nm and c = 1.71 nm. The form of optical interest is called *Iceland Spar*, and is a particularly clear form of the material. Iceland Spar crystals are easily cleaved into rhombohedra.

If such a rhombohedron is placed over a black spot on a sheet of paper *two* lines or spots will be seen on looking from above through the crystal, as shown in colour Plate 3.1 and schematically in Figure 3.6. The spots also appear to be at different heights within the crystal itself. One spot will appear to be undeviated in position with respect to the spot on the paper. The undeviated spot is formed by light moving through the crystal as if it were glass, and the ray producing this effect is variously called the *ordinary ray*, the *O-ray* or *o-ray*. If the crystal is then rotated the "ordinary" spot will remain in place while the other will rotate in a circle about the fixed spot. The ray causing this behaviour is called the *extraordinary ray*, *E-ray* or *e-ray*. The crystal is displaying the fact that it has two indices of refraction and the feature is called *double refraction*.

If a sheet of Polaroid is placed over the crystal and rotated, at first one dot disappears and then the other. This is illustrated in Figure 3.6e, f. If the crystal is picked up and tilted, the separation of the two dots will change, and if it is possible to look down the diagonals of the rhombohedron, in one case only one dot will be seen, that formed by the o-ray, no matter how the crystal is rotated about this diagonal. This direction is called the *optic axis*. In a normal cleaved rhobohedron of calcite the optic axis lies along the body diagonal which passes through the "bluntest" pair of corners. These occur at the two corners where the faces which meet all show obtuse angles.

In fact the same effect will be observed with all tetragonal, hexagonal and trigonal crystals but unless the two refractive indices are quite different the effect is too small to be noticed casually. All crystals in these systems will have one optic axis, which is the crystallographic c-axis. These crystals are called *uniaxial* in mineralogical texts.

Similar effects will be seen with crystals belonging to the orthorhombic, monoclinic and triclinic systems, except that these have two optic axes. These crystals are referred to as *biaxial* in mineralogical texts.

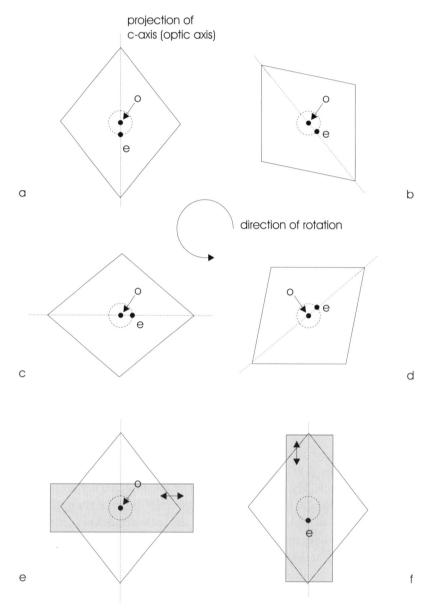

Figure 3.6 Schematic representation of the appearance of double refraction by a crystal of calcite placed over a black spot on a sheet of paper. (a–d) As the crystal is rotated one image (due to the o-ray) remains stationary and one (due to the e-ray) rotates. (e, f) A sheet of Polaroid placed with its vibration direction, indicated as a double headed arrow, perpendicular to the projection of the c-axis (the optic axis) of the calcite causes the e-ray to disappear, while the same Polaroid rotated by 90° causes the o-ray to disappear

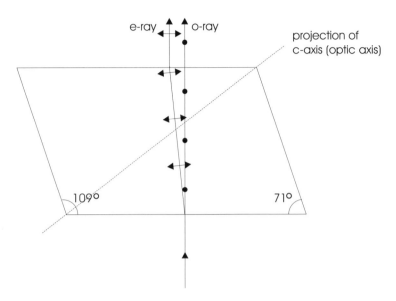

Figure 3.7 The passage of a monochromatic beam of light through a cleaved prism of calcite. Normal light falling perpendicularly upon the lower face of the prism is split into two components with different polarisation. The o-ray, with a vibration direction in a plane perpendicular to the c-axis, indicated by filled circles along the ray, is undeviated. The e-ray, with a vibration direction in a plane which includes the c-axis, indicated by double headed arrows, is deviated by about 6°. The top and bottom cleavage faces of the prism are (101) planes with respect to the hexagonal unit cell and the c-axis (the optic axis) is a body diagonal of the cleavage rhombohedron

3.6 THE EXPLANATION OF DOUBLE REFRACTION EFFECTS

The cause of the phenomena just described can be attributed to the behaviour of the crystal with respect to the polarisation of the light. To return to calcite as an example, a beam of normal light traversing the crystal is split into two, as drawn in Figure 3.7. The o-ray is undeviated and consists of light linearly polarised with vibration direction parallel to the base of the crystal rhomb, indicated as filled circles along the ray. The e-ray is deviated and consists of light linearly polarised with the vibration direction perpendicular to that in the o-beam, indicated by double headed arrows along the ray. No light is absorbed and half of the incident intensity is found in each of the beams.

When unpolarised light is transmitted along the c-axis (the optic axis) the polarisation is unimportant and all the light behaves as if the refractive

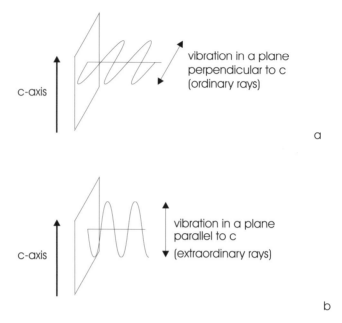

Figure 3.8 Unpolarised light incident normal to the optic axis is resolved into two beams which are linearly polarised. (a) Linearly polarised light with a vibration direction in a plane perpendicular to the c-axis forms the ordinary, o-ray or O-ray. (b) Linearly polarised light with a vibration direction in a plane parallel to and containing the c-axis forms the extraordinary, e-ray or E-ray

index was n_o = 1.658. When unpolarised light is transmitted in any other direction, the crystal will show two refractive indices. One of these will always be equal to n_o but the other one will depend upon the direction of the light ray and so be variable, written n'_e. When the unpolarised light is transmitted perpendicular to the c-axis (the optic axis) it is resolved into two beams, one with a vibration direction perpendicular to c, the o-ray, and one with a vibration direction parallel to c, the e-ray. This is sketched in Figure 3.8. The variable refractive index n'_e is now found to be at a maximum or minimum, and the resulting refractive index, simply called n_e, is 1.486. The relationship between the magnitude of n'_e and the angle θ that the ray makes with the optic axis is:

$$1/(n'_e)^2 = \cos^2\theta / n_o^2 + \sin^2\theta / n_e^2$$

When the light beam travels parallel to c the value of θ is zero and n'_e is equal to n_o and there is only one effective refractive index for the material.

When the light beam travels perpendicular to c the value of θ is 90° and n'_e is equal to n_e. The values of the refractive indices n_o and n_e are called the *principal indices* of the crystal. The difference between the principal indices is called the *birefringence* of the crystal:

$$n_o - n_e \text{ or } n_e - n_o \text{ for a uniaxial crystal}$$

$$n\gamma - n\alpha \text{ for a biaxial crystal.}$$

where $n\gamma$ and $n\alpha$ are the greatest and smallest refractive indices of the crystal.

Calcite has n_o = 1.658, n'_e = 1.486 up to 1.658, n_e = 1.486. The birefringence is $n_o - n_e$ = (1.658 − 1.486) = 0.172.

To summarise, in *all* crystals of symmetry lower than cubic the refractive index depends upon the direction of vibration of the light ray. Any ray not passing down an optic axis is resolved into two rays linearly polarised in two mutually perpendicular directions.

3.7 COLOUR PRODUCED BY POLARISATION AND DISPERSION

The production of a spectrum by a prism of glass (or any isotropic substance) has been described in Chapter 2. If the prism is made of a uniaxial or biaxial material two spectra can form, as depicted in Figure 3.9. In Figure 3.9a a prism made of a uniaxial material is drawn with the optic axis perpendicular to the paper. The incident beam of unpolarised light is split into two, an ordinary and extraordinary beam, each plane polarised, as described above. Each of these will produce a spectrum. If the prism is cut so that the optic axis is in the plane of the paper, Figure 3.9b, only one spectrum will form.

3.8 PLEOCHROISM AND DICHROISM

If uniaxial or biaxial crystals are viewed by transmitted linearly polarised white light, many will be seen to change colour on rotation. This feature is called *pleochroism* (many coloured). Uniaxial crystals may display two colours (*dichroism*) and biaxial crystals three colours (*trichroism*). There are many examples of pleochroism that could be cited. Plates of the

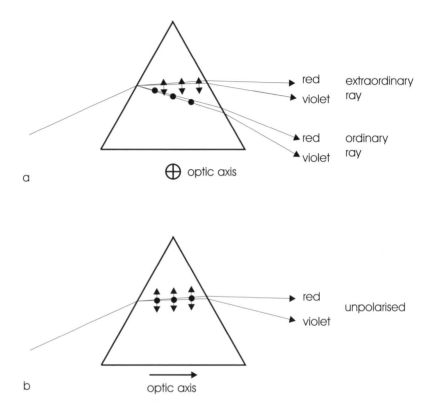

Figure 3.9 (a) A prism made of a doubly refracting material will produce two spectra with normal white light when the optic axis is perpendicular to the beam. The extraordinary ray is polarised parallel to the optic axis (shown as double-headed arrows) and the ordinary ray is polarised perpendicular to the optic axis (shown as filled circles). (b) When normal light is propagated along the optical axis only one spectrum forms as the ordinary and extraordinary rays are not separated

trigonal mineral tourmaline about 1 mm thick and containing the c-axis will transmit light with a vibration direction parallel to the c-axis. If the crystal is rotated through 90° the crystal will become dark as light polarised perpendicular to the c-axis is almost completely absorbed, as in Figure 3.10. This is a frequently used way of producing linearly polarised light in an optical microscope.

Ruby (corundum, Al_2O_3, containing about 0.5% chromium oxide, Cr_2O_3) belongs to the hexagonal system and is dichroic. If viewed in linearly polarised white light with the plane of vibration parallel to the c-axis (the optic axis) the crystal appears orange-red. When rotated by 90° so that the

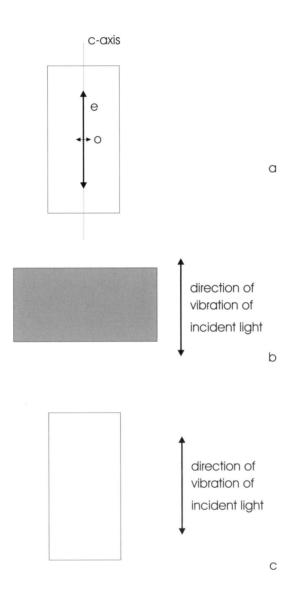

Figure 3.10 (a) A plate of the mineral tourmaline, cut so as to contain the crystallographic c-axis (the optic axis), transmits linearly polarised light differently depending upon the direction of vibration. The relative transmission factors are shown by the orthogonal pair of double headed arrows. (b) An observer positioned above the crystal which is illuminated from below with linearly polarised light will see either a dark crystal (b) or a clear crystal (c), depending upon the orientation of the slab with respect to the direction of vibration of the light

plane of vibration is perpendicular to the optic axis the colour seen is purple-red. The absorption spectra of ruby for both of these situations are reproduced in Figure 7.12.

The reason for pleochromism is that the absorption of light is dependent both upon its direction in the crystal with respect to the optic axes and its state of polarisation. In these materials the absorption spectrum of a crystal with light polarised along one crystal axis is different to the absorption spectrum when light is polarised along the other axes.

3.9 COLOUR PRODUCED BY POLARISATION AND INTERFERENCE

It is known that two beams of light which are polarised with mutually perpendicular vibration directions never interfere. However, interference can occur if a beam of linearly polarised light is split into two by passage through an anisotropic crystal and then the vibration planes of these beams are made to coincide by passage through another polar.

Many beautiful colours can be seen in thin plates of crystals in this way. For example, if a thin sheet of calcite is viewed by transmitted white light when sandwiched between crossed polars an interference colour will be seen. The incident plane polarised beam will be split into two and a path difference will be introduced in each beam, given by (dn_o) for the ordinary ray and (dn'_e) for the extraordinary ray, where d is the real thickness of the plate. The relative difference in path between these rays, the *retardation*, Δ, is given by $d(n_o - n'_e)$. On passing through the analyser the two beams will be arranged so that the vibration directions are parallel and the path difference will result in interference and the production of colour. A sequence of colours will be seen which will vary as a function of the thickness of the plate and its orientation with respect to the optic axis.

The production of colour is not limited to plates cut carefully with respect to the optical axes. Generally, any thin sheet of an anisotropic material will suffice. A good example is provided by a sheet of mica placed between two polars and viewed in transmission by holding the sandwich up to a white light. Suppose that the polars are crossed. As the mica sheet is rotated with respect to the polars four positions, at 90°, will be found at which the mica sheet becomes dark. These are the *extinction* positions. When the mica sheet is midway between these positions it will seem to be brightly coloured. The colour seen is very sensitive to the viewing angle, but if care is taken to look at the foil without any change of viewing angle the colour will be seen to remain unchanged as the mica sheet is rotated. It is only the overall intensity which changes.

The colour is due to the retardation produced because the mica sheet displays two indices of refraction. As only the absolute difference between the refractive indices is important, the retardation is given by:

$$\Delta = d|n_1 - n_2|$$

where n_1 and n_2 are the apparent refractive indices of the plate and the positive value of the difference is taken. The variation in colour with viewing angle noted above occurs because the value of d, the thickness of the mica sheet traversed by the light rays, changes with viewing angle. The colour observed as a function of retardation is given in Appendix 4.1. If the crossed polars are now rotated to be in the parallel position without changing the orientation of the mica, the complementary colour will be seen, as set out in Appendix 4.1. The colour seen when the mica is between crossed polars is thus the colour subtracted from the white light when the polars are in a parallel orientation. More detailed descriptions of the fascinating colours observed when crystals interact with polarised light will be found in some of the texts listed in the Further Reading section at the end of this chapter.

Surprisingly this same effect can be exploited to reveal stress and strain in an isotropic material. (The result of a stress (a force or load applied to a material) is a strain (a deformation).) In the previous chapter the relationship between density and refractive index was described. When a material is stressed the density will change slightly. If the stress is directional the density will vary in a pattern which mirrors this. Thus an isotropic material under stress can contain optically anisotropic regions. If the material is observed between crossed polars coloured fringes will reveal the stressed areas. In effect the stress encodes information on the linearly polarised incident beam which is decoded by the analyser. The effect is easily seen. Take a piece of plastic film and look at it between crossed polars. Generally nothing of interest will be seen. If you now stretch the film (technically subject it to a uniaxial tensile stress) brightly coloured areas will appear. The birefringence so produced in the now anisotropic film is called *stress birefringence* and the effect is known as *photo-elasticity*. This feature is widely used in glass blowing to make sure that residual stress is not present. A glass workpiece is viewed between crossed polars and those regions that are stressed are revealed. If necessary the piece can then be reheated at a moderate temperature (*annealed*) to allow the glass to flow slightly and so relieve the stress present. Before the advent of high speed computers the stress in complex engineering components could be analysed by building them of clear plastic and viewing the stress and strain fields present using crossed polars. Regions of the structure containing high levels of stress show coloured fringes, the spacing of which indicates the stress gradients present. Colour Plate 3.2 shows a thin piece of polymer film used to wrap food (cling-film) stretched and viewed between crossed polars. The bright colours in the normally transparent film reveal regions of high strain in the film.

This phenomenon is also well known to car drivers who wear Polaroid sunglasses. The windscreens of cars are stressed in a predetermined way so as to avoid catastrophic failure if hit by a flying stone or similar object. Light reflected from a hot road will be partly polarised, as explained above. The Polaroid sunglasses act as an analyser and coloured fringes delineating the stressed areas are clearly visible over the windscreen.

3.10 NON-LINEAR EFFECTS

Imagine a bright green beam of light emerging from a 1 cm cube of a perfectly transparent crystal, with no apparent electrical or other connections to it. What is happening is that a beam of invisible laser radiation in the infrared is being converted by the crystal into a beam of green light. There are a number of ways that this can come about. Here we will consider only pure undoped crystals and generally refer to the phenomenon as *frequency doubling* or *second harmonic generation* (SHG). *Up-conversion*, another way of adding two photons together so as to produce a photon of doubled frequency, is discussed in Section 9.8. In up-conversion an impurity ion acts so as to achieve the optical transformation. (The language of the literature is often less than precise and frequently doubling is often called up-conversion.) Frequency doubling utilises only the crystal matrix and impurity dopants are not involved. Frequency doubling is a *non-linear effect*.

This effect comes about in the following way. In Chapter 2 we saw that the interaction of light with transparent materials involves electronic polarisation by the electric field of the light wave. For light beams of ordinary intensity the polarisation, P, is a *linear* function of the electric field, E:

$$P = \varepsilon_0 \chi E$$

where ε_0 is the permittivity of free space and χ is the dielectric susceptibility of the material, which is proportional to the relative permittivity and
refractive index of the substance. However this is only an approximation
and in the general case it is better to write the polarisation as a series:

$$P = \varepsilon_0\chi^{(1)}E + \varepsilon_0\chi^{(2)}E^2 + \varepsilon_0\chi^{(3)}E^3 + \ldots \quad (3.1)$$

where $\chi^{(1)}$ is the linear dielectric susceptibility, $\chi^{(2)}$ the second-order
dielectric susceptibility, $\chi^{(3)}$ the third-order dielectric susceptibility and so
on. The polarisation is no longer a simple linear function of the electric
field.

The extra "non-linear" terms are unimportant for all materials when
ordinary light is considered. The advent of lasers changed this and sufficiently high electric fields have become available for the second and third
terms in the series to take on a significance. These additional terms are only
high enough to be of importance in relatively few materials and in general
the magnitude of the values of the dielectric susceptibilities decreases
rapidly as the order increases, so that the second-order, $\chi^{(2)}$, term is the
most important *non-linear coefficient*. However, the symmetry of the
crystal is important and for centrosymmetric materials only the odd order
terms have non-zero values. In non-centrosymmetric crystals, which lack a
centre of symmetry, the second-order term, $\chi^{(2)}$, is present and it is these
types of material that are generally known as *non-linear optical materials*.

The non-linear terms in the polarisation equation allow photons to be
added and subtracted in certain specific ways. For example, if the crystal is
irradiated with laser light characterised by *angular frequencies* ω_1 and ω_2 a
collection of frequencies $2\omega_1$, $2\omega_2$, $\omega_1+\omega_2$ and $\omega_1-\omega_2$ can all be produced. It
is the production of a frequency $2\omega_1$, the *second harmonic*, from a single
input frequency ω_1, that is known as frequency doubling or second
harmonic generation. It is depicted schematically in Figure 3.11a. The
production of the other frequencies, schematically illustrated in Figure
3.11b, is known as *frequency mixing*.

Second harmonic generation comes about in this way. The electric field
associated with a light beam varies in a sinusoidal fashion and can be
written:

$$E = E_0\cos \omega t$$

where ω is the angular frequency of the light. If this is substituted into
Equation 3.1 it is found that:

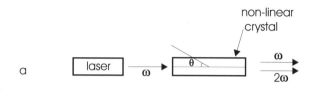

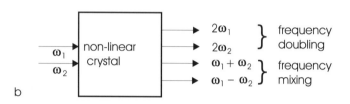

Figure 3.11 (a) A schematic illustration of the production of frequency doubled light using a non-linear crystal set with its optical axis at the phase matching angle, θ, to the input beam. This is also called second harmonic generation, SHG. As the incident beam traverses the crystal it is gradually transformed into the frequency doubled component. (b) Frequency doubling and frequency mixing with a non-linear crystal

$$P = \omega_0\chi^{(1)}\,E_0\cos\omega t + \varepsilon_0\chi^{(2)}\,(E_0\cos\omega t)^2 + \varepsilon_0\chi^{(3)}\,(E_0\cos\omega t)^3 + \ldots$$

This series can be written more simply as:

$$P = A + B\cos\omega t + C\cos 2\omega t + D\cos 3\omega t \ldots$$

where A, B, C and D are constants. At field strengths of the order of those found in laser light the wave which emerges from the crystal has both the ω and the higher 2ω, 3ω and higher angular frequencies present. Remember, however, that $\chi^{(2)}$ is zero in centrosymmetric crystals, which makes them useless for second harmonic generation. In fact the production of a second harmonic from a crystal when illuminated by a laser is usually taken as a good test for the lack of a centre of symmetry.

If two input waves are present we can understand how frequency mixing comes about. Suppose the crystal is irradiated with two beams simultaneously.

$$E_1 = E_{01}\cos\omega_1 t$$

$$E_2 = E_{02}\cos\omega_2 t$$

The electric field in the sample is then:

$$E = E_1 + E_2 = E_{01}\cos \omega_1 t + E_{02}\cos \omega_2 t$$

Substituting this into Equation 3.1 will produce a series for P which will contain a term

$$\tfrac{1}{2}[\cos (\omega_1 + \omega_2)t + \cos(\omega_1 - \omega_2)t]$$

This is able to give rise to two output waves, one of frequency $(\omega_1 + \omega_2)$ and one of frequency $(\omega_1 - \omega_2)$.

3.11   FREQUENCY MATCHING AND PHASE MATCHING

In principle any non-centrosymmetric crystal can be used for the generation of other frequencies in agreement with the description just given. As the incident beams traverse the crystal they are gradually converted from one angular frequency to the others, such that:

$$\omega_1 + \omega_2 = \omega_3 \qquad\qquad (3.2)$$

This equation sets out the *frequency matching* condition for frequency mixing.

This creates a problem if it is desired to obtain a reasonably intense output of the newly created waves as the refractive index of the material depends upon wavelength. Recall that:

$\omega = 2\pi\nu = 2\pi c/\lambda$
n = velocity of light in vacuum (c) / velocity of light in crystal (v)
  $= \lambda(\text{vacuum}) / \lambda(\text{crystal})$

The consequence of this is that the newly formed beams within the crystal will be out of phase with each other and the incident beam. As we have seen, beams with a phase difference can interfere with each other. A result of this interference is that the intensity of the new rays emerging from the crystal will be very low. Now the destructive interference can be prevented if all of the beams remain in phase. In the crystals that we are speaking of this can be achieved by making the refractive indices and angular frequencies agree with the equation:

$$n_1\omega_1 + n_2\omega_2 = n_3\omega_3 \qquad (3.3)$$

This is known as the *phase matching* condition, which can be compared with the frequency matching condition given by Equation 3.2.

Although obtaining *phase matching* would appear to be a tall order, it can be achieved in a number of uniaxial crystals. Consider second harmonic generation. It is necessary to satisfy Equations 3.2 and 3.3 in the case where

$$\omega_1 = \omega_2 = \omega; \; \omega_3 = 2\omega$$

i.e.

$$\omega + \omega = 2\omega$$

and

$$n\omega + n\omega = n'2\omega$$

The trick is to treat the initial beam, angular frequency ω, as the ordinary ray and the frequency doubled beam, angular frequency 2ω, as the extraordinary ray in a uniaxial crystal. It is then necessary to find a crystal direction, in which the refractive index for the ordinary ray, n_o, with an angular frequency of ω, has the same refractive index, n'_e, as the extraordinary ray with an angular frequency of 2ω. Now this will not be possible in many crystals, but it is possible in some.

In suitable crystals one angle with respect to the optic axis will be found, the *phase matching angle*, which can be used. For crystals in which n_e is greater than n_o, known as *positive uniaxial crystals*, a useful approximate relationship is:

$$\sin^2\theta \sim [n_o(\lambda) - n_o(2\lambda)/\Delta n$$

where θ is the phase matching angle, n_o the refractive index of the ordinary ray at wavelengths λ and 2λ and Δn is the birefringence, $n_e - n_o$. The equation indicates that the phase matching angle is determined by the ratio of the dispersion to the birefringence. When this angle is known the crystal is accurately aligned so that the laser beam exactly follows the correct path. At this point destructive interference no longer occurs as all the beams remain in step. In this case the frequency doubled intensity can become considerable. For example, in the material $ZnGeP_2$, which can be used for second harmonic generation with CO_2 lasers to produce radiation in the 5 μm range from a primary output in the 11 μm range, the phase matching angle is 53.9° to the optic axis.

Some other commercially available non-linear optical crystals suitable for second harmonic generation and frequency mixing are ammonium dihydrogen phosphate, $(NH_4)H_2PO_4$ (ADP), potassium dihydrogen phosphate, KH_2PO_4 (KDP), barium bromate, $BaBrO_3$, lithium iodate, $LiIO_3$, barium borate, BaB_2O_4 (BBO), lithium triborate, LiB_3O_5 (LBO), potassium niobate, $KNbO_3$ and sodium barium niobate, $NaBa_2Nb_5O_{15}$. The *conversion efficiency* of these crystals, that is, the amount of incident light converted to the higher frequency, is of the order of 10–15 per cent.

Because the refractive index is temperature sensitive, crystals have to placed in temperature controlled cells to achieve reasonable amounts of conversion. In addition, for optimum efficiency it is important to eliminate out surface reflections at the entry face of the crystal. This is achieved by coating this face with an antireflection layer. How these coatings are achieved is described in the next chapter.

3.12 OPTICAL PARAMETRIC AMPLIFIERS AND OSCILLATORS

An *optical parametric amplifier* is a device which uses a non-linear crystal to amplify a wave. The foundation of the method is *three-wave mixing* in which

$$\omega_1 + \omega_2 = \omega_3$$

In an amplifier, a *signal* wave, of angular frequency ω_1 which has a low intensity, is passed through a non-linear crystal. At the same time a very intense wave, of angular frequency ω_3, called the *pump* wave, is also passed through the crystal in the same direction. In suitable crystals, the pump wave is gradually decomposed into two waves with angular frequencies ω_1 and ω_2 as it crosses the crystal. The newly formed wave of angular frequency ω_1 adds to, and so amplifies, the signal wave. The wave of angular frequency ω_2 is redundant and is called the *idler* wave.

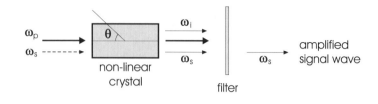

a

b

Figure 3.12 (a) A schematic illustration of the operation of an optical parametric amplifier in which the crystal transforms the pump wave (ω_p) plus the signal wave (ω_s) into an amplified signal wave plus a redundant idler wave (ω_i). (b) The arrangement of an optical parametric oscillator (OPO) in which the crystal transforms the pump wave (ω_p) alone into an amplified signal wave (ω_s) and a redundant idler wave ($_i$). The optical axis of the crystal employed has to be set at an angle near to the phase matching angle, θ, for efficient conversion and output

In effect the pump wave is split into two, in agreement with the frequency matching relationship:

$$\omega_p = \omega_s + \omega_i$$

For a reasonable output the crystal must be aligned to achieve phase matching for all frequencies. The equation of importance is:

$$n_p\omega_p = n_s\omega_s + n_i\omega_i$$

where the subscripts refer to the pump, signal and idler waves and n is the relevant refractive index. In a practical device a filter is used to remove the unwanted frequencies from the output beam, as sketched in Figure 3.12a.

Optical parametric oscillators (OPOs) are similar but do not require initial input of a signal wave. As described above, the pump wave produces waves with angular frequencies ω_1 and ω_2. If the crystal is arranged between two mirrors which can reflect these waves, one completely and one almost completely, these will continually bounce to and fro through the crystal. A little will escape through the imperfect mirror. (We shall see how to manufacture such mirrors out of thin films in the following chapter.) The pump wave, passing through the crystal, will gradually be converted into the signal and idler components as before, and an amplified signal wave can be obtained as output. This process is depicted in Figure 3.12b.

In a uniaxial crystal the refractive indices encountered by the pump, signal and idler waves are a continuous function of the angle of the waves to the optic axis. This means that there are a range of allowed combinations of phase matching and frequency matching conditions which can occur. The actual frequencies generated, ω_s and ω_i, will therefore vary as the angle of the incident beam on the non-linear crystal is changed. This is known as *tuning*.

The effect is considerable. For example a commercial oscillator, using a barium borate (BBO) crystal and 532 nm pump radiation, can produce a signal wave varying in wavelength from 650 to 1060 nm and an idler wave varying in wavelength from 1060 to 3000 nm by rotation of the crystal over an angle of just 2°. The limitations to the useful output arise from the range of transparency of the crystals and the range over which the crystals can be rotated while still maintaining reasonable phase matching.

3.13 LIQUID CRYSTALS

Early experiments established that normal liquids were isotropic and had no effect upon the polarisation of light. Concurrent with this, it was observed that some organic crystals derived from cholesterol seemed to have two "melting points". For example, cholesteryl benzoate appeared to show a lower "melting point" (when the crystals turned into a cloudy liquid), and an upper "melting point" (when the liquid became clear), separated by a temperature interval of 33 K.

The cloudy region was described as consisting of one or more meso-morphic phases or, more usually now, *mesophases*. When studied with a polarising microscope the mesophase region, although certainly liquid-like, had a noticeable effect upon polarised light and seemed to behave rather like a low symmetry crystal. Because of this, these curious materials were referred to as *liquid crystals*.

A century or more of investigation has shown that the liquid crystals first investigated are made up of needle-like rigid molecules. On raising the temperature a liquid crystal disorders in a number of steps rather than all

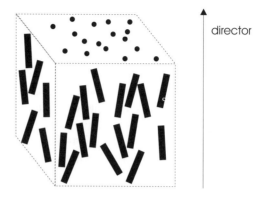

Figure 3.13 A schematic illustration of a nematic structure. The centres of the elongated rigid molecules (black bars) which make up the phase are arranged at random, but the molecules themselves tend to preferentially align along a direction called the director, which is vertical in the diagram

at once as is normal for the solid–liquid transition. Above the first "melting point" the molecules lose the strictly ordered intermolecular spacing typical of a normal crystal but still retain a partial degree of order. One of the commonest forms of disorder is found when the molecules retain a roughly parallel orientation but the geometric centres of the molecules are arranged at random as in a conventional liquid. This state of affairs is shown in Figure 3.13. The preferred direction along which the molecules align is called the *director*. Such a situation occurs in a *nematic liquid crystal* (from the Greek, nematos, thread), so called because when viewed in polarised light long dark threads appear to occur throughout the bulk. These dark thread-like lines are optical effects caused by linear defects called *disclinations* which run through the structure. Most nematic liquid crystal mesophases have only one optical axis and so are uniaxial.

The structures which form within the mesophase region are complex and depend upon intermolecular forces and temperature as well as the geometry of the components. Moreover, the orientation of the molecules is easily influenced by external disturbances such as electric fields. At higher temperatures, above the second "melting point", the molecules become truly random in direction and a normal liquid forms with no effect on polarised light.

Recently it has been found that liquid crystal-like behaviour can be obtained from materials which are built up from disc-shaped molecules. To differentiate the two types, rod-like molecular liquid crystals are called *calamitic* (from the Greek, kalamos, reed) while disc-like molecular liquid crystals are called *discotic*.

Because nematic liquid crystals behave as uniaxial materials they can generate colours in polarised light in the same way as uniaxial crystals. In addition, the fact that the nematic arrangement interacts with an electric field enables these materials to be widely used in liquid crystal displays. These are described in Chapter 12. Moreover, some forms of liquid crystals, called *cholesteric* or *twisted nematic* phases, can produce colour directly by diffraction. This aspect is described in Chapter 6.

Finally, we can remark that many of these molecules themselves show nonlinear effects because they lack a centre of symmetry. Currently there is a great deal of research taking place on the use of organic molecules as nonlinear materials.

3.14 OPTICAL ACTIVITY

One of the most intriguing results obtained by scientists trying to unravel the physics and chemistry of living materials during the last century was the phenomenon of *optical activity*. For example, crystals of salts of the two acids *tartaric acid* and *racemic acid* were well known and could be collected from old wine casks. The sodium salts of these two acids, sodium tartrate and sodium racemate, were chemically identical. However, if linearly polarised light was passed through a solution of the tartaric acid salt the plane of polarisation rotated to the right. The amount of rotation, the *specific rotation*, was as good a physical property of the compound as, for instance, melting point, and it could be used for characterisation purposes. The puzzle was that the corresponding salt of racemic acid was optically inactive and caused no rotation.

The resolution of the problem was glimpsed when Louis Pasteur made a painstaking optical examination of sodium racemate crystals. In 1848 he announced that these contained equal numbers of two forms, one "right-handed" and one "left-handed", meaning that they had the same relationship to each other as a left-hand glove to a right-hand glove or an object and its mirror image. Solutions of the two crystal types rotated the plane of polarisation by equal amounts but in opposite directions.

The process of dissolution separates the solid crystals into molecules or ions. It was clear, therefore, that at least this example of optical activity was a feature which needed to be explained at a molecular level and not entirely at the crystallographic level. Since then it has long been

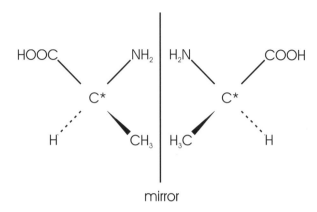

Figure 3.14 Schematic diagram of the enantiomers of the amino acid alanine. The chiral carbon atom in each molecule is marked C*. The chemical bonds formed by these atoms have a tetrahedral geometry. Bonds in the plane of the paper are represented as lines, bonds projecting out of the plane of the paper towards the reader are drawn as heavy triangles and those projecting below the plane of the paper downwards are represented by dotted lines

established that any molecule that can exist in two mirror image related forms is optically active. They are referred to as *chiral* molecules. The pair of mirror image molecules are called *enantiomers* and in organic chemistry are also known as *optical isomers*. In organic compounds optical isomers occur whenever four different groups are attached to a tetrahedrally co-ordinated central carbon atom, making it a *chiral carbon atom* or *chiral centre*. The two enantiomers of the amino acid alanine, in which the chiral carbon atom is marked C*, are shown in Figure 3.14. Although it is not easy to see from drawings that the two structures cannot be superimposed, the construction of a simple model will convince you. Tartaric acid, Pasteur's crystals, are more complex because the molecules contain two chiral carbon atoms. These can "cancel out" internally in the molecule so that three molecular forms actually exist, the two optically active mirror image structures which cannot be superimposed on each other (referred to as laevorotatory and dextrorotatory, see below) and the optically inactive form, which can be superimposed on its mirror image. These are drawn in Figure 3.15. Once again, simple models greatly help in understanding the fundamental difference between these molecules.

Inorganic molecules with tetrahedral or octahedral bond geometry can also form optically active pairs. Many optically active crystals contain optically active molecules, but this is not mandatory. Crystals of quartz,

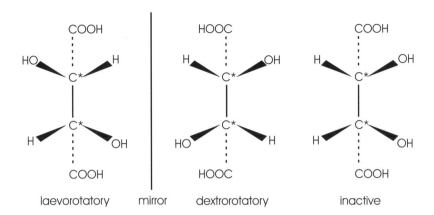

COOH   HOOC   COOH

laevorotatory   mirror   dextrorotatory   inactive

Figure 3.15 Schematic diagram of the three forms of tartaric acid. Two of these, the laevorotatory and dextrorotatory forms, are mirror images and are optically active. The third form is inactive. The chiral carbon atoms in each molecule are marked C*. The chemical bonds formed by these atoms have a tetrahedral geometry. Bonds in the plane of the paper are represented as lines, bonds projecting out of the plane of the paper towards the reader are drawn as heavy triangles and those projecting below the plane of the paper downwards are represented by dotted lines

one form of silicon dioxide, SiO_2, occur in left- or right-handed forms although no molecules are present.

Enantiomers have identical chemical properties except when they react with other chiral molecules. This has profound effects for life because most biologically important molecules are chiral. The differences in biological and pharmacological activity between the two enantiomers can be pronounced. One enantiomer can trigger a sensation of the odour of caraway while the other enantiomer that of spearmint. Vitamin C prevents the disease scurvy, the other enantiomer of this optically active substance is biologically inactive. Such a list could be extended indefinitely. It has been known for a number of years that living creatures make use of left-handed amino acids and right-handed sugars. The reason for this has attracted numerous answers, some rather fantastic, but it now seems that the bias is not restricted to life on Earth. Studies of the Murchison meteorite which were reported in 1997, indicated that it too contains left-handed amino acids.

Enantiomers, of course, differ in one physical property: they display *optical activity*. One form of a chiral molecule will rotate the plane of polarised light in one direction and its enantiomer will rotate it in the opposite direction. The form of molecule which rotates the plane of polarisation to the right is labelled *dextrorotatory*. The form of molecule which rotates the plane of polarisation to the left is called *laevorotatory*. Mixtures of enantiomers in equal proportions will not rotate the plane of linearly polarised light and are called *racemic* mixtures, after the "racemic acid" of Pasteur.

Optically active compounds are not coloured by virtue of this property. However, the optical activity is often translated into colours when these compounds are viewed between polars. This is due to interference between the rotated and non-rotated components of the light beam after they have passed through the analyser. The effect is similar to that discussed in Section 3.9 with respect to thin crystal plates. Colour generation using the principles of rotation of polarised light in displays is explained in Chapter 12.

3.15 ANSWERS TO INTRODUCTORY QUESTIONS

Why do some crystals produce double images?
Some crystals produce double images because the index of refraction is different in different directions in the crystal. (In fact this is true of all crystal systems except cubic.) The result of this is that unpolarised light incident on the crystal is divided into two components unless the ray follows a particular crystal lattice direction (an optic axis). This forms the two images. The phenomenon is called double refraction.

Why do the colours of some crystals depend upon the viewing direction?
The colours are subtraction colours after the light has passed through the crystal and so it is necessary to consider the light absorbed. If the incident light is polarised the light absorbed will depend upon the angle between the plane of polarisation and the crystal optic axes. Rotation of the crystal will change this angle and so change the absorption spectrum. The colour transmitted will change because of this. The effect is called pleochroism.

How can some prisms produce either one spectrum or two?
If a beam of white light shines on a crystal (apart from cubic materials) in a direction perpendicular to the optic axis it will divide into two rays of differing polarisation. Each of these rays will give rise to a spectrum because of the dispersion of the prism material. If the prism is cut so that the incident beam enters parallel to the optic axis the splitting does not occur and only one spectrum forms.

How can infrared radiation of wavelength 1.08 μm be changed into green light?

This can be carried out using frequency doubling. For this it is necessary to have a low symmetry crystal. To obtain a reasonably intense output it is necessary to propagate the light rays in a particular direction in the crystal.

What are liquid crystals?
Liquid crystals are phases with a structure intermediate between that of a crystalline solid and an isotropic liquid. They are frequently composed of rod-like molecules which are partly ordered to form a mesophase. The degree of ordering is lost at the upper melting point of the mesophase, at which temperature an ordinary liquid forms in which there is no order between the molecules.

3.16 FURTHER READING

Polarisation from the viewpoint of optics is described by:

F. A. Jenkins and H. E. White, Fundamentals of Optics, 3rd edition, McGraw-Hill, New York (1957).

O. S. Heavens and R. W. Ditchburn, Insight into Optics, Wiley, New York (1991).

A. Nussbaum and R. A. Phillips, Contemporary Optics for Scientists and Engineers, Prentice-Hall, Englewood Cliffs, NJ (1976).

B. E. A. Saleh and M. C. Teich, Fundamentals of Photonics, Wiley, New York (1991).

A collection of classic papers on polarised light, which includes reprints of studies by Huygens and Newton, and which makes fascinating reading is:

W. Swindell, ed. Polarised Light, Dowden, Hutchison and Ross, Pennsylvania (1975), distributed by John Wiley.

Crystallography is covered in detail in:

J. V. Smith, Geometrical and Structural Crystallography, Wiley, New York (1982).

The relationship between crystal properties and light is summarised by:

F. D. Bloss, Crystallography and Crystal Chemistry, Holt Rinehart and Winston, New York (1971), Ch. 11.

R. E. Stoiber and S. A. Morse, Crystal Identification with the Polarising Microscope, Chapman and Hall, London (1994).

E. E. Wahlstrom, Optical Crystallography, 5th edition, Wiley, New York (1975).

Some information on non-linear optical materials is to be found in:

O. S. Heavens and R. W. Ditchburn, Insight into Optics, Wiley, New York (1991), Ch. 15.

A. Nussbaum and R. A. Phillips, Contemporary Optics for Scientists and Engineers, Prentice-Hall, Englewood Cliffs, NJ (1976), Ch. 14.

B. E. A. Saleh and M. C. Teich, Fundamentals of Photonics, Wiley, New York (1991).

Information on liquid crystals is given in:

P. J. Collins, Liquid Crystals: Nature's Delicate Phase of Matter. Princeton University Press, Princeton (1990).

Optical activity is discussed in all textbooks concerned with organic chemistry and many concerned with inorganic chemistry. See, for example:

K. P. C. Vollhardt and N. E. Schore, Organic Chemistry, 3rd edition, W. H. Freeman, San Francisco (1999).

D. F. Schriver, P. W. Atkins and C. H. Langford, Inorganic Chemistry, 2nd edition, Oxford University Press, Oxford (1994).

3.17 PROBLEMS AND EXERCISES

1. Estimate the angle at which light reflected from water will be completely polarised.

2. Light is reflected from a polished glass plate with n = 1.563 at an angle of 15°. Calculate the reflectance for the s- and p-wave.

3. What is the intensity of light transmitted by a polariser–analyser combination with their vibration directions arranged at 45°?

4. Make sketches of the shapes of the unit cells for the seven crystal systems.

5. Calculate the birefringence and sketch the unit cells of the minerals scheelite, $CaWO_4$, tetragonal, n_α = 1.920, n_β = 1.936; corundum, Al_2O_3, hexagonal, n_α = 1.761, n_β = 1.769; forsterite, Mg_2SiO_4, orthorhombic, n_α = 1.635, n_β = 1.651, n_γ = 1.670; malachite, $Cu_2(OH)_2CO_3$, monoclinic, n_α = 1.655, n_β = 1.875, n_γ = 1.907; sterconite, $Na(NH_4)H(PO_4).4H_2O$, triclinic, n_α = 1.439, n_β = 1.442; n_γ = 1.469.

6. Determine the (variable) refractive index n'_e for quartz for a ray travelling at 20° to the optic axis. Quartz is hexagonal and the c-axis is

the optical axis. $n_o = 1.544$, $n_e = 1.553$. What does the calculation for an angle of 45° tell you?

7. A mica foil is viewed in transmitted white light between crossed polars approximately midway between extinction conditions. It appears orange-red-brown. When the polars are rotated to the parallel position the colour changes to blue-green. Estimate the retardation using Appendix 4.1. If the refractive indices of the mica are taken as 1.602 and 1.563 calculate the thickness of the foil.

8. Show that the conditions described in Section 3.12 for the optical parametric oscillator barium borate (BBO) to produce a signal wave varying from 650 to 1060 nm and an idler wave varying from 1060 to 300 nm from 532 nm pump radiation agree with the formula

$$\omega_{pump} = \omega_{signal} + \omega_{idler}$$

CHAPTER 4

THE PRODUCTION OF COLOUR BY REFLECTION

Why are soap bubbles coloured?
Why are thin films of oil on water coloured?
How are antireflection coatings on lenses produced?
How can perfect mirrors be made from transparent
materials?

Reflection is a commonplace phenomenon. Nevertheless, it can give rise to a surprising range of colours. The most vivid of these are associated with the presence of reflection by a thin transparent film. Bright colours are often seen in soap bubbles and close examination of transparent insect wings shows that these can show areas which are beautifully coloured. Casual observation also reveals that the colours seem to vary with the direction of viewing and with the thickness of the film. "Metallic" and opalescent car paints and cosmetics are recent examples of similar effects. In this chapter the origin of these and other colours due to reflection is explored.

4.1 REFLECTION FROM A POLISHED SURFACE

When light falls onto a transparent surface such as a glass plate some of it will be reflected, as shown in Figure 4.1.

For reflection:

$$\theta_i = \theta_r$$

where θ_i is the *angle of incidence* and θ_r the *angle of reflection*. The plane of incidence contains the incident ray, the reflected ray and the normal to the reflecting surface. In Figure 4.1 this is the plane of the page.

The amount of light reflected from a surface such as glass depends upon the polarisation of the light, discussed in Chapter 3. For a polished plate

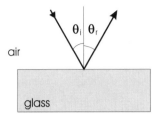

Figure 4.1 Light incident on a glass plate will be reflected. The angle of incidence, θ_i, will be equal to the angle of reflection, θ_r

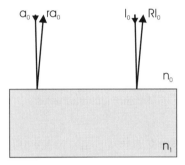

Figure 4.2 Reflection of a beam of light from a transparent surface. The amplitude of the incident beam is a_0 and incident intensity is I_0. The reflected amplitude will be given by ra_0 and the reflected intensity by RI_0. The angles have been exaggerated for clarity. The formulae in the text assume that the light is perpendicular to the surface

and light at *normal incidence* (i.e. perpendicular to the surface), the polarisation can be ignored and the *coefficient of reflection*, r, is given by:

$$r = (n_0 - n_1)/(n_0 + n_1)$$

where n_0 and n_1 are the refractive indices of the media on the two sides of the boundary in the direction in which the light travels, as indicated in Figure 4.2. The coefficient of reflection is defined such that if a wave of *amplitude* a_0 falls upon the surface, then the *amplitude* of the reflected wave is ra_0. For reflection at a surface between a substance of low refractive index and a substance of high refractive index, r is negative. This signifies a phase change of π radians on reflection, which means, in terms of a light wave, that a peak turns into a trough and vice versa, as illustrated in Figure 4.3.

The eye detects *intensity* changes rather than amplitude changes, and so it is the more convenient to work with the *reflectivity* or *reflectance*, R:

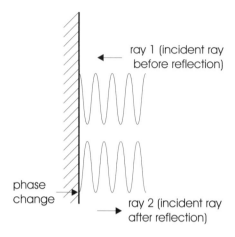

Figure 4.3 Reflection introduces a phase change of $\lambda/2$ in a ray reflected at a surface of higher refractive index

$$R = r^2 = [(n_0 - n_1)/(n_0 + n_1)]^2$$

This is because the intensity, I_0, is proportional to the square of the amplitude, $(a_0)^2$. The reflected intensity, $R(I_0)$, (see Figure 4.2) is then proportional to $r^2(a_0)^2$.

The reflectivity, R, for a glass plate of refractive index n in air is:

$$R = [(1 - n)/(1 + n)]^2$$

$$= [(n - 1)/(n + 1)]^2 \qquad (4.1)$$

Remember that because n depends upon wavelength, the reflectivity will vary across the spectrum.

For ordinary glass R is about 0.03–0.04, that is, about 3–4 per cent. Although this may seem to be a rather insignificant amount, it is noticeable in everyday life. Reflections from windows and from the glass over a painting are frequently annoying. Moreover, it is too high for specialist purposes such as high performance lenses and optical components designed for use in laser systems, so these are given antireflection coatings, discussed below. However, even this small degree of reflectivity turns out to be an essential component in the production of colour through interference by thin films, as we now consider.

4.2 INTERFERENCE AT A SINGLE THIN FILM IN AIR

Suppose monochromatic light travelling through air falls upon a homogeneous thin film of refractive index n, as sketched in Figure 4.4a. Light will be reflected from the top surface to give the reflected ray 2. The light transmitted into the film will be repeatedly reflected from the bottom surface and the underside of the top surface as shown. At each reflection some of the light will escape to produce additional reflected and transmitted rays. As the reflectivity is rather small the first reflected ray and the first transmitted ray are of most importance.

Let us first consider the situation occurring when a light ray falls onto the surface at perpendicular incidence, as in Figure 4.4b. Some of the reflected light seen by the observer will have been reflected at the top surface (ray 2). However, some will have travelled through the film and been reflected from the bottom surface before reaching the observer (ray 3). Because of the difference in the paths taken by the two rays, the waves comprising these rays will be out of step. In addition, as mentioned above, because ray 2 is travelling through a medium of low refractive index and is reflected at a surface of higher refractive index a wave peak will turn into a trough and vice versa. This will not happen to ray 3 because it is travelling in the high refractive index material and is reflected at the high/low surface. The combined effect of these changes is drawn in Figure 4.4c. It is convenient to refer to the difference in the positions of corresponding points on the two waves associated with rays 2 and 3 as the *retardation*. (The term retardation is widely used in mineralogy to refer to the relative "slowing down" of one ray with respect to another as they both pass through a crystal. In the present case it is synonymous with the path difference due to all of the interactions with the thin film.) In this condition interference between the two waves can now occur, which will cause the film to look either dark or bright.

The effect is easily understood. Ray 3 will have travelled further than ray 2 by twice the film thickness, which we will call d. Now we are interested in how much the light waves in rays 2 and 3 are out of step and it is necessary to use the optical path, $[d]$, given by:

$$[d] = nd$$

The optical path difference between rays 2 and 3 will be $[p]$, where:

$$[p] = 2[d] = 2nd$$

where d is the physical thickness and n is the refractive index of the film.

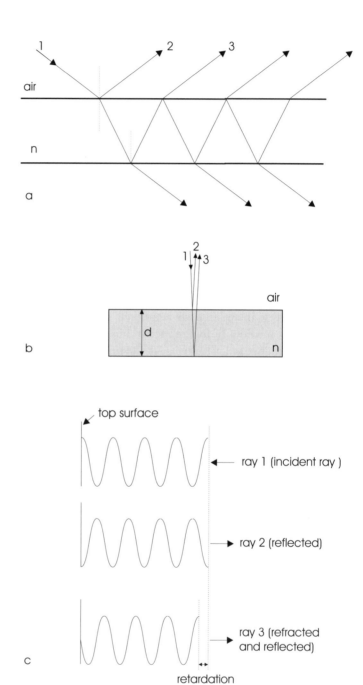

We now have all the information needed to specify the appearance of a thin film when viewed by reflection at normal incidence. If the path difference, [p], considered in isolation, is equal to an integral number of wavelengths the waves will be exactly in step at the surface. Adding in the phase change of exactly half a wavelength for ray 2 will make it out of step with ray 3 by this amount as they leave the surface. The film will therefore appear dark due to destructive interference. This is summarised by:

$$[p] = m\lambda \quad (m = 1,2,3, \dots) \quad \text{gives a } minimum \text{ (dark)}$$

Acting alone, a path difference, [p], between rays 2 and 3 equal to a half-integral number of wavelengths will cause the two rays to be exactly out of step. Adding in the half wavelength phase change for ray 2 will make them exactly in step. The film will then appear bright, because constructive interference will occur. In summary:

$$[p] = (m + 1/2)\lambda \quad (m = 1,2,3, \dots) \quad \text{gives a } maximum \text{ (bright)}$$

At other path differences the film will appear to have an intermediate tone, depending upon the exact difference between rays 2 and 3.

4.3 THE VARIATION OF INTERFERENCE WITH VIEWING ANGLE

The previous section described the situation occurring when a beam of light falls perpendicularly onto a single thin film in air. Here we consider what happens if the light beam is at an angle to the surface. The geometry of the situation is drawn in Figure 4.5. In the case illustrated, θ is the angle of incidence and θ' the angle of refraction. The path difference, [p], between rays 2 and 3 is found to be:

Figure 4.4 (a) The reflection and transmission of a ray of light incident on a transparent film in air. There will be a number of reflected and transmitted beams due to repeated reflection at the top and bottom faces of the film. Interference can occur between the reflected components and between the transmitted components. (b) At normal incidence (the angles of incidence and reflection have changed from 90° for clarity) ray 3 will have travelled further than ray 2 by an optical path difference of 2nd. (c) The waves making up rays 2 and 3 will be out of step due to the combined effects of a phase change on reflection and the path difference by an amount called the retardation

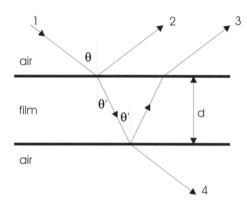

Figure 4.5 The geometry of reflection at a thin film in air. The incident ray 1 is partly reflected to give ray 2 and partly transmitted and reflected from the bottom surface of the film to give ray 3. The remaining part of the ray is transmitted out of the film as ray 4. Interference between rays 2 and 3 is possible due to the path difference introduced in the reflection

$$[p] = 2nd \cos \theta'$$

The analysis now follows that given in the previous section. If [p] is equal to a whole number of wavelengths the film will appear dark, due to the combined effect of path difference and change of phase of ray 2 on reflection at the surface. Thus

$$[p] = 2nd \cos \theta' = m\lambda \quad \text{gives a } \textit{minimum} \text{ (dark)}$$

For the same reason, if the path difference turns out to be a half wavelength reinforcement will occur and we find:

$$[p] = 2nd \cos \theta' = (m+1/2)\lambda \quad \text{gives a } \textit{maximum} \text{ (bright)}$$

This indicates that when viewed at an angle in monochromatic light some regions of the film will look bright and others dark. As the viewing angle changes so will the areas which appear dark and light. If the light is normal to the film, $\cos \theta = \cos \theta' = \cos (0) = 1$ and the equations reduce to those given in the previous section.

Plate 1.1 A moorland scene displaying colours due to scattering (the blue sky) and absorption (the green-browns of grass and soil). The blue of the stream is simply reflection of blue light and is not caused by the interaction of white light with the water

Plate 2.1 A halo-like arc of colour produced by reflection and dispersion when sunlight falls upon hexangonal prismatic ice crystals in the upper atmosphere

Plate 3.1 Double diffraction by a rhombohedron of Iceland Spar. If the crystal is rotated the separation of the two pairs of lines visible through the crystal will alter, and in some orientations a pair may merge to appear as a single line

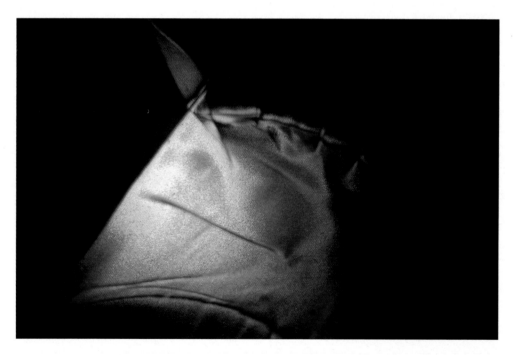

Plate 3.2 A thin piece of polymer film used to wrap food (cling-film) stretched and viewed between crossed polars. The bright colours in the normally transparent film reveal regions of high strain in the film

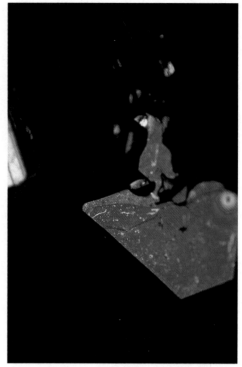

Plate 4.1 Interference of colours due to white light reflected from thin transparent plates of molybdite (molybdenum trioxide, MoO_3). The differences in colour are due to variation in crystal thickness

Plate 4.2 Colours due to white light interference in a thin transparent film of silicon dioxide (SiO_2), on carborundum (silicon carbide, SiC). The colour variation is due to changes in film thickness

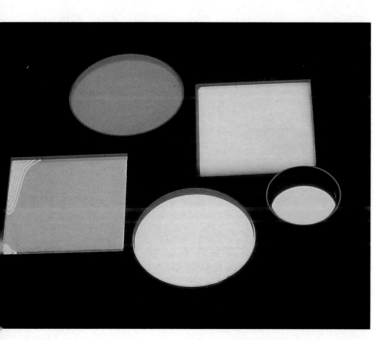

Plate 4.3 Multilayer interference filters. The bright colours reflected are complementary to the colours transmitted by the filters and absorbed by the black backing

Plate 4.4 (a) The blue butterfly *Polyommatus icarus*. The blue colour arises from multiple thin film reflection from colourless scales cloaking the upper side of the wings. Figure reproduced by the kind permission of Dr J.A. Findlay

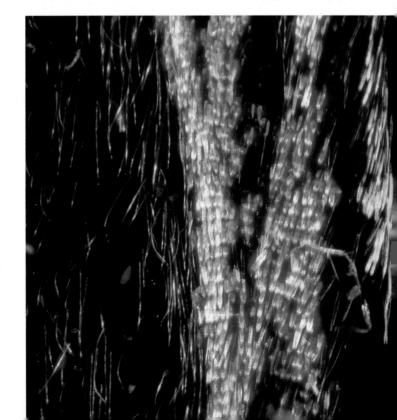

(b) Individual scales from a wing of *Polyommatus icarus*. Only some scales have a blue reflecting microstructure. The yellow-brown scales contain melanin-related pigments

4.4 THE COLOUR OF A SINGLE THIN FILM IN AIR

Although a discussion of monochromatic light is helpful so as to understand the physical processes taking place on reflection at a thin film we are really much more interested in what the appearance of the film will be in daylight. When the film is viewed in white light, the same reflection and interference discussed above will occur, except that we have to take into account the effects of all of the different wavelengths present.

This idea is illustrated in Figure 4.6. For example, violet light with a wavelength of 400 nm will reflect a minimum of intensity when the film produces a path difference, [p], or retardation of $m\lambda$, i.e. 400 nm, 800 nm, 1200 nm and so on. This is shown in Figure 4.6a. The situation for the wavelengths 450 nm, 500 nm, 550 nm, 600 nm, 650 nm and 700 nm is also drawn.

In order to determine the reflected colour of a thin film when viewed in white light it is necessary to add up all of these contributions over all of the values of the retardation. (In reality a continuum of wavelengths occurs between 400 nm and 700 nm. Here we have taken seven wavelengths to simplify things.) Take, for example, a film of thickness equivalent to a retardation of 600 nm, represented by the dotted line in Figure 4.6. It is seen that there is a large contribution from the 400 nm wave. The contribution from succeeding waves decreases until at a wavelength 600 nm there is no contribution at all. Thereafter a small contribution is obtained from wavelengths of 650 and 700 nm. The overall colour perceived will be the sum of all of these. Because of the dominance of the 400 nm contribution the film will appear to be a violet-blue colour.

The intensity pattern and perceived colour generated by adding up all of the contributions as a function of retardation is shown in Figure 4.7. The sequence of colours seen will repeat in a cyclical fashion as the film thickness increases or decreases, as certain colours are either reinforced or cancelled. Each sequence of spectral colours is called an *order*, which starts with the *first order* for the thinnest of films. A new order begins every 550 nm of retardation. The appearance of a thin film viewed by reflection in white light is given in Appendix 4.1. These interference effects are illustrated in Plate 4.1, which shows thin crystals of molybdite (molybdenum trioxide, MoO_3) viewed in reflected white light.

Ultimately, strong interference effects will be lost. This is because ordinary white light is emitted in bursts which undergo a sudden change of phase every 10^{-8} s or so. When the film is very thin the two rays which interfere come from within the same burst and interference effects are

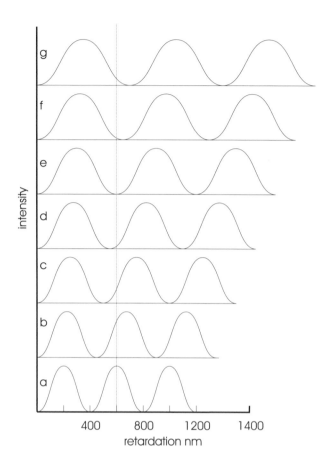

Figure 4.6 The intensity (in arbitrary units) of light reflected from a single thin film in air at various wavelengths plotted as a function of the retardation. (a) λ = 400 nm; (b) λ = 450 nm; (c) λ = 500 nm; (d) λ = 550 nm; (e) λ = 600 nm; (f) λ = 650 nm; (g) λ = 700 nm

noticeable. With thicker films interference takes place between different bursts of light and the interference effects are weaker. Initially this results in films showing mainly pale pinks and greens in the fourth and fifth orders. With even thicker films all interference effects are smoothed out and colours are no longer apparent to the eye.

The properties so far described refer to the reflected light. Consider now the transmitted light rays shown in Figure 4.4a. Since the fraction of incident white light which is reflected is coloured it follows that the transmitted light will be depleted in this colour. The colour seen will

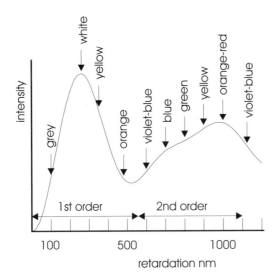

Figure 4.7 The total intensity (in arbitrary units) reflected from a thin film in air illuminated by white light as a function of retardation. The colours observed by eye are indicated

therefore be the *complementary colour* to that strongly reflected. This is listed in Appendix 4.1.

If the angle of viewing is not perpendicular to the film the retardation changes slightly. The correct retardation is given by:

$$[p] = 2[d] \cos \theta'$$

where θ' is the angle of *refraction* as set out in Section 4.3. (But note that at all angles except for perpendicular viewing, polarisation will also occur and be important.) This formula indicates that as the viewing angle moves away from perpendicular to the film the colour observed will move towards lower retardation. Figure 4.7 reveals, for example, that second order orange-red will change towards green and blue.

This discussion explains the familiar colours of soap films seen in air. These are best seen if the film is viewed against a black background, which prevents the effects being masked by other reflections. As the thickness of the films vary, due to water flow within the films themselves, the colours seen change in a dramatic and beautiful way. If the films are on a wire frame, the transmitted and reflected colours can be compared.

4.5 THE REFLECTIVITY OF A SINGLE THIN FILM IN AIR

Interference and colour, as just discussed, should be differentiated from reflectivity. It could be that a certain colour, say red, is produced by interference effects in a film, but whether the colour is readily seen will depend upon the reflectivity of the film for this wavelength. The reflectivity of a thin film in air will be different from that for a thick plate, given in Equation 4.1, as interference effects from the bottom surface also need to be considered. Additionally, the polarisation of the light will be important and can only be neglected when the light is incident normal to the surface of the film.

For light at normal incidence the reflectivity is given by a rather complex formula:

$$R = 2r_1^2 - 2r_1^2\cos2\delta \; / \; 1 - 2r_1^2\cos 2\delta + r_1^4$$

where $r_1 = (n_0 - n_f \; / \; n_0 + n_f)$; n_0 is the refractive index of the surrounding medium, usually air, with $n_0 = 1.0$, n_f is the refractive index of the film, and

$$\delta = 2\pi[d] \; / \; \lambda$$

[d] is the optical thickness of the film, given by:

$$[d] = n_1 d$$

and d is the physical thickness. The reflectivity is found to vary in a cyclic fashion from zero for values of [d] equal to 0, $\lambda/2$, λ, etc., to a maximum (of 0.15 for $n_f = 1.5$) for values of [d] given by $\lambda/4$, $3\lambda/4$ and so on. Because the value of n_f depends on wavelength, the reflectivity will also vary across the spectrum.

4.6 THE COLOUR OF A SINGLE THIN FILM ON A SUBSTRATE

The behaviour of a single thin film on a substrate is similar to that discussed for the case of a single thin film in air. Now, however, it is necessary to take into account any change of phase that might occur on reflection at the back surface of the film. If the substrate has a *lower refractive index* than the film on the surface then the treatment will be identical to that for a thin film in air, from the point of view of interference effects. If the refractive index of the substrate is greater than that of the film then a phase change will be introduced at both the air–film interface and the film–substrate interface.

If a thin film on a substrate is viewed in white light colours will be seen. The actual hue perceived will be found by a summation of all of the reflected intensities, as was discussed earlier.

In the case of a thin film on a substrate of lower refractive index, the reflected colours observed when the film is viewed at normal incidence in white light will be the same as those listed in the "colour reflected" column of the table in Appendix 4.1. (The transmitted colours are normally absorbed by the substrate.) If the thin film is on a substrate of *higher refractive index* than the film itself a change of phase will now take place at the lower boundary as well as at the top surface. In this case the reflected colour seen at normal incidence when viewed in white light will be the *complementary* colour to that just described. These are also listed in Appendix 4.1 in the column labelled "colour transmitted". For example, if it is necessary to estimate the thickness of a film of SiO_2 grown on the surface of a single crystal of silicon by observing its colour in reflected white light, it is necessary to use the "colour transmitted" list because the refractive index of silicon is greater than that of silica.

The colour produced by a thin film on a substrate is shown in Plate 4.2, which illustrates colours due to interference in a thin film of transparent silicon dioxide (SiO_2) on a crystal of carborundum (silicon carbide, *sic*). When first grown the crystals of carborundum are a shiny black. However, they soon take on a wide variety of attractive iridescent colours because of surface oxidation which produces thin silicon dioxide surface films in a wide variety of thickness.

Colours produced in this way are also frequently seen when a thin oil film covers a puddle of water on a road. These colours are enhanced to the eye by the black road surface, which absorbs all light not reflected by the film. As the thickness of the oil changes in response to wind or water movement the colours vary considerably.

4.7 THE REFLECTIVITY OF A SINGLE THIN FILM ON A SUBSTRATE

The reflectivity of a single thin film deposited on a substrate, like that of a single thin film in air, depends upon the polarisation of the light, the film thickness and direction of the incident radiation. In the case of illumination involving light of one wavelength perpendicular to a homogeneous non-absorbing thin film:

$$R = (r_1^2 + 2r_1r_2\cos2\delta + r_2^2) / [1 + 2r_1r_2\cos2\delta + r_1^2r_2^2)$$

where $r_1 = (n_0 - n_f) / (n_0 + n_f)$; $r_2 = (n_f - n_s) / (n_f + n_s)$, n_0 is the refractive index of the surrounding medium, n_f the refractive index of the film and n_s the refractive index of the substrate. The expression for δ is:

$$\delta = 2\pi[d] / \lambda$$

[d] is the optical thickness of the film, given by:

$$[d] = n_f d$$

and d is the physical thickness of the film. For values of [d] given by

$$[d] = \lambda/2, \lambda, 3\lambda/2, \text{etc.}$$

the equation reduces to:

$$R = (n_0 - n_s)^2/(n_0 + n_s)^2$$

This is *identical* to the equation that we found for an uncoated surface. Thus a layer of optical thickness $\lambda/2$, etc., can be considered to be *optically absent* and the surface has normal uncoated reflectivity. This is an intriguing and useful result. It means that if a delicate surface is coated with a $\lambda/2$ layer of a hard transparent material the surface will be protected without any effect on optical properties!

When the optical thickness of the film is:

$$[d] = \lambda/4, 3\lambda/4, \text{etc.}$$

the reflectivity is given by:

$$R = [(n_f^2 - n_0n_s)/(n_f^2 + n_0n_s)]^2 \tag{4.2}$$

and the reflectance will be *either a maximum or a minimum*. This will depend upon whether the film has a higher refractive index than the substrate or a lower refractive index than the substrate. When the film has a *higher* refractive index than the substrate the reflectivity will be a *maximum* (more explicitly, when $n_0 < n_f > n_s$). When the film has a *lower* refractive index than the substrate the reflectivity will be a *minimum* (more explicitly, when $n_0 < n_f < n_s$).

As with a thin film in air, the value of the reflectivity will cycle with film thickness between a lower value at [d] equal to 0, $\lambda/2$, λ, etc., to a maximum for values of [d] equal to $\lambda/4$, $3\lambda/4$ and so on. Because the value of n_f depends on wavelength, the reflectivity will also vary across the spectrum.

4.8 Low-reflection (Antireflection) and High-reflection Coatings

We can easily use the above equations to see how thin films modify the reflectivity of a surface. Suppose that it is desired to make a non-reflective coating on a glass surface in air. (Such coatings are called antireflection (AR) coatings.) Turning to Equation 4.2 we find that if the value of n_f lies between that of air and the glass the reflectivity will be a *minimum* for a $\lambda/4$ film. Putting R equal to zero in Equation 4.2 yields a value of the refractive index of the film which will give no reflection at all:

$$n_f = \sqrt{(n_s)} \qquad (4.3)$$

For glass n_s is about 1.5 so the antireflecting film must have a refractive index:

$$n_f = \sqrt{(1.5)} = 1.225$$

Very few solids have such a low index of refraction, and a compromise material often used is magnesium fluoride, MgF_2, for which n in the middle of the visible is 1.384. This is not perfect, but does reduce the reflectivity from about 4% down to about 1%. The coating will actually be maximally antireflective for the *design wavelength*, which is the wavelength for which the calculations were made. For camera lenses, which commonly use antireflection coatings, the design wavelength is usually near the middle of the visible spectrum, say 550 nm. Such films reflect violet and red more than yellow or green; an effect readily visible when a good coated camera lens is examined. The reflectance of a quarter wavelength single film antireflection coating of MgF_2 on a glass surface (n = 1.52) when the beam is perpendicular to the surface (angle of incidence 0°) is illustrated in Figure 4.8. The film shown has a design wavelength of 550 nm.

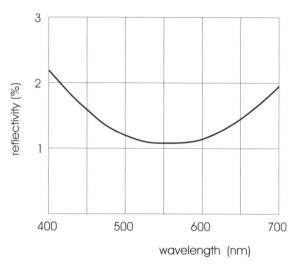

Figure 4.8 The reflectivity of a quarter-wave thick film of a magnesium fluoride (MgF$_2$) antireflection coating on a glass surface with a refractive index of 1.52. The design wavelength of the film is 550 nm and the beam is taken as perpendicular to the surface

To optimise the reflectivity and make the value of R closer to 1 one can also use a film of thickness λ/4, but in this case n$_f$ should be greater than n$_0$ and n$_s$. Two materials frequently used are silicon monoxide SiO$_x$ with x approximately equal to 1.0 (n = 2.0) or titanium dioxide TiO$_2$ (n = 2.90). A TiO$_2$ film of thickness λ/4 on glass will have a reflectivity of about 0.48 (48%). As R for a single glass surface in air is about 0.04 (4%), almost 48% represents a great improvement. The effect is used in costume jewellery. Rhinestones consist of a glass object with n close to 1.52 coated with an approximately λ/4 thickness film of TiO$_2$. Variations in film thickness and viewing angle give these objects a wide variety of fleeting colours which are meant to simulate the fire of diamonds. Sparkling paints and nail varnish also make use of an approximately quarter wavelength thickness of TiO$_2$ deposited onto flakes of mica which are subsequently dispersed in the product. The various colours seen are created in a similar way to the colours on rhinestones.

4.9 ANTIREFLECTION LAYERS

Apart from their utility as surface coatings, antireflection (AR) layers are also becoming important in the manufacture of semiconductor components. The fabrication of an integrated circuit on a silicon chip involves one or more steps in which the material is exposed to light through a pattern called a mask. The mask is used to selectively illuminate areas on the chip which, after further processing, build into the array of transistors which manipulate data. The light actually interacts with a layer of substance called a photoresist. After illumination the photoresist is weakened in those areas which were exposed to light and these are subsequently dissolved away so as to reveal the underlying silicon, which can then be selectively doped or otherwise treated. The length of time of the exposure of the photoresist to light is critical to the success of the process.

The desire to pack more and more transistors onto a chip has led to the drawing of finer and finer detail onto the mask and the use of shorter and shorter wavelength light in the illumination steps. At present the use of ultraviolet radiation is commonplace. The sharpness of the pattern produced on the silicon and hence the number of transistors which can be placed onto the chip is limited by diffraction (see Chapter 6) and multiple reflections within the photoresist. The multiple reflections expose parts of the photoresist which should remain unexposed, as illustrated schematically in Figure 4.9a. This has the effect of reducing the sharpness of the projected pattern and can also introduce spurious detail or defects.

In order to combat this difficulty an antireflection layer can be applied between the silicon substrate and the photoresist, as shown in Figure 4.9b. The aim is to introduce a film of the correct thickness to ensure that the reflected rays passing through the photoresist are out of phase by $\lambda/2$ so that destructive interference occurs in the photoresist layer. Although the idea is conceptually simple the thickness of the antireflection layer is rather difficult to determine.

There are two main reasons for this. First, as the layer is interposed between the silicon and the photoresist the simple formula (Equation 4.3) given in Section 4.8 cannot be applied and rather complex calculations of the reflectivity must be made. Second, the familiar refractive index term, n, of the layer must be replaced by a more complex function. This is because at wavelengths in the ultraviolet region many materials which are transparent at visible wavelengths absorb strongly. In order to take absorption into account the refractive index, n, must be replaced by the complex refractive index N where:

$$N = n + ik$$

In this equation k is variously termed the *extinction coefficient, coefficient of absorption* or *attenuation coefficient* and i represents the complex term $\sqrt{-1}$. Note that both n and k vary with wavelength.

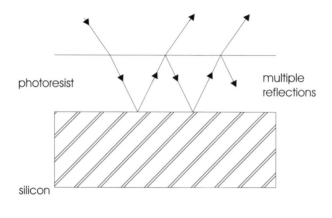

Figure 4.9 (a) Multiple light reflections in a film of photoresist on a silicon surface. (b) The deposition of an antireflection layer between the photoresist and the silicon results in cancellation of reflected beams by destructive interference, thus increasing the precision of the process

One suitable material that has been used in antireflection layers is silicon oxynitride, SiO_xN_y, often erroneously written as SiON in technical reports. The compound has an advantage in that a change of composition alters the optical properties of the film. Some values of the refractive index and extinction coefficient are given in Table 4.1. The material is laid down as a thin film by passing a mixture of silane, SiH_4, nitrous oxide, N_2O, and nitrogen, N_2, over the silicon wafer. The various proportions of the gases

Table 4.1 Optical properties of Si–O–N films at 248 nm
wavelength as a function of composition

Composition	Refractive index, n	Extinction coefficient, k
$SiO_{0.86}N_{0.24}$	1.8948	0.4558
$SiO_{0.71}N_{0.27}$	1.9682	0.5253
$SiO_{0.54}N_{0.59}$	2.0821	0.5004
$SiO_{0.47}N_{0.49}$	2.2127	0.6030

controls the values of x and y in the film, which allows the wavelength at
which the film is optimally antireflective to be varied at will. The SiO_xN_y
layer is thus said to be a *tuneable* antireflection layer.

4.10 DIELECTRIC MIRRORS

Traditionally mirrors have been made from metals. The best metallic
mirrors are made of a thick layer of silver, which has a reflectivity of about
0.96. (The reflectivity of metals is considered in more detail in Section
10.3.) Surprisingly, multiple thin films of transparent materials can be laid
down one on top of the other in such a way as to form perfect mirrors.
These are often called *dielectric mirrors*.

The simplest formulae for the reflectance of such a mirror refers to the
specific case in which all layers are λ/4 thick and of alternating high (H)
and low (L) refractive indices, n_H and n_L, illuminated by light falling *perpendicular* to the surface. The arrangement, drawn in Figure 4.10, is called
a *quarter-wave stack*. We will consider a quarter-wave stack deposited on
a substrate in the sequence:

substrate; L; H; L; H; L; H; . . . L; H; air

Maximum reflectance is given by the formula:

$$R = [(n_sf - n_0) / (n_sf + n_0)]^2$$

where f is equal to $(n_H/n_L)^{2N}$, n_0 is the refractive index of the surrounding
medium, usually air ($n_0 = 1.0$), n_s is the refractive index of the substrate,
usually glass ($n_s \sim 1.5$) and N is the number of (LH) *pairs* of layers in the
stack.

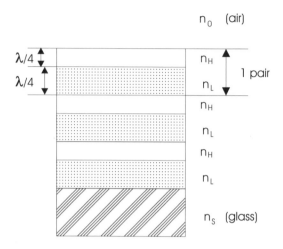

Figure 4.10 A stack of thin films, each of optical thickness $\lambda/4$, called a quarter-wave stack, can act as an effective dielectric mirror

This is equivalent to:

$$R = [(n_s - (n_L/n_H)^{2N}) / (n_s + (n_L/n_H)^{2N})]^2$$

for a stack in air. Computation shows that as the number of pairs of layers increases R rapidly approaches 1, implying perfect reflectivity. Note that different formulae must be used if the stack terminates with an L-layer, if we do not have complete sets of pairs of layers or if oblique illumination is considered.

Dielectric mirrors are now routinely used in precision optical work. They are made of layers of relatively common materials including silica glass and a few oxides and fluorides. These are all stable in air and the mirrors have the additional advantage over metallic mirrors of not degrading in normal use.

4.11 MULTIPLE THIN FILMS AND PHOTONIC ENGINEERING

The general approach used to make a dielectric mirror is to lay down a stack of thin films which have alternately higher and lower refractive indices. Manipulation of the thickness' and the refractive indices of the layers in the stack allows one to modify the optical properties at will and to produce far more than simply high reflectivity. The products of this technology which are readily available include, as well as virtually perfect mirrors, virtually perfect antireflection coatings, both of which can be tuned to respond to very specific wavelengths, and a variety of optical filters. The fabrication of such devices is known as *photonic* or *thin film engineering*.

Filters utilising multilayers, sometimes referred to as *interference filters*, fall into three different categories. *Shortpass filters* transmit visible wavelengths and cut out infrared radiation. They are often used in surveillance cameras to eliminate heat radiation. A schematic illustration of the transmittance of such a filter is shown in Figure 4.11a. *Longpass filters*, with a transmission profile similar to that sketched in Figure 4.11b, block ultraviolet radiation and transmit the visible. Other filters, called *bandpass filters*, pass only a limited section (or band) of the electromagnetic spectrum. A typical transmittance profile for a bandpass filter is drawn in Figure 4.11c. (Note that these thin film interference filters generally give a much sharper transmittance than the type of filter made from dye molecules distributed in a gelatine matrix; the type of filter illustrated in Chapter 1, Figure 1.5). As the filters are made of transparent layers the wavelengths not transmitted are reflected. Bandpass filters therefore act as mirrors for the complementary colour of the transmitted band. Because of this effect, these filters are often vividly coloured. A collection of multilayer interference filters is shown in colour Plate 4.3.

The reason why such devices are rather recent in origin is that even the slightest change in the number of layers used, the thickness of any layer or the relative refractive indices produces a considerable change in the optical properties. This is coupled with added complexity which arises when the angle of incidence of the incident light or its degree of polarisation changes. This has meant that satisfactory multilayer devices which perform to design standards have only been available with the advent of sophisticated computer programmes which allow the computations needed to be carried out reasonably quickly and accurately. To illustrate this, Figure 4.12 shows the variation in reflectivity of a stack of four thin films as the thickness of just one of the layers is changed. The four thin films are deposited on a glass substrate and alternate between high refractive index, (H), and low refractive index, (L) ending with air. The arrangement of the layers is:

air (n = 1.0)
L, 93 nm, n = 1.48
H, 120 nm, n = 2.30

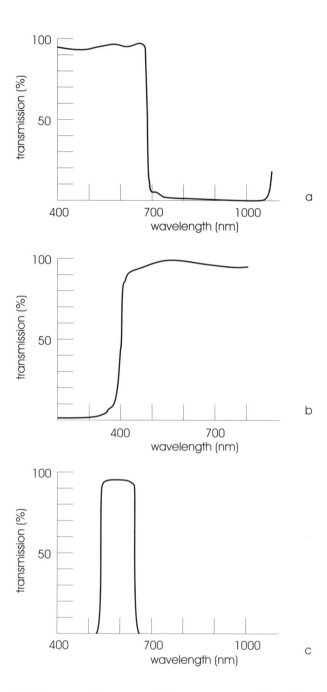

Figure 4.11 Schematic illustration of the transmission profiles of (a) a shortpass filter, (b) a longpass filter and (c) a bandpass filter

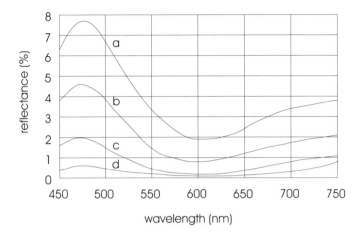

Figure 4.12 The reflectivity of a stack of four thin films on a glass substrate in air. The thickness of each layer is constant except for the one adjacent to the glass, which takes values of (a) 30 nm, (b) 24 nm, (c) 18 nm and (d) 12 nm. The stack (d) shows almost perfect antireflection behaviour. Computations were made using "Filmstar" software (see Section 4.15)

L, 37 nm, n = 1.48

H, variable thickness, 30 nm, 24 nm, 18 nm, 12 nm, n = 2.30

substrate, glass, n = 1.52

The single layer to be changed was that next to the glass substrate, and then only from a thickness of 30 nm to 12 nm. The curves are all evaluated for a design wavelength of 550 nm and for light normally incident upon the stack. The final curve, Figure 4.12d, makes an almost optimal antireflection coating. (The curves were derived using free software generously supplied by F. Goldstein, listed in Section 4.15.) Thanks to such software, computations are now routinely possible for stacks of films containing dozens of separate layers.

Despite the undoubted complexity of thin film optics there is one general result that is useful. When multiple thin film mirrors or filters are tilted the wavelength which is strongly reflected or transmitted shifts towards lower values, but it does so much more slowly than for a single thin film, as described above. For example, a 30° tilt will change a strong reflection at 550 nm to one at about 530 nm which is far smaller than would be experienced if a single thin film were tilted by the same amount.

Figure 4.13 A scanning electron micrograph of the internal structure of a blue scale from the wing of the butterfly *Polyommatus icarus* (L). The scale is transparent and the blue colour arises from reflection and interference within the multiple thin film arrangement in the scale interior, clearly visible in the micrograph

4.12 PHOTONIC ENGINEERING IN NATURE: THE COLOUR OF BLUE BUTTERFLIES

Plate 4.4a is a photograph of a blue butterfly. The wings of butterflies are covered with minute scales, which can be seen in Plate 4.4b. In the case of blue butterflies the scales giving rise to the blue colour are actually transparent and do not contain blue pigments. Instead, they consist of a multilayer arrangement of lamellae which is similar to that just discussed, as can be seen from the electron micrograph in Figure 4.13. Each scale is made up of several layers of transparent material with a refractive index of about 1.5 separated by air layers of the same thickness. The thickness of each layer is close to 100 nm, producing a reasonable approximation of a quarter-wave stack with reference to violet or blue light. The arrangement is highly reflecting for wavelengths towards the blue region of the spectrum. The blue reflection is enhanced by the presence of black scales which lie under the transparent scales and so absorb all of the light not reflected. This form of optical engineering has been perfected by many years of evolution to give the species comprising this family of insects many beautiful and varied blue colours. Indeed, some of these butterflies have developed scales which are made up of quarter-wave stacks which strongly reflect ultraviolet wavelengths. Although these reflections are not visible to us they are to other butterflies, whose vision extends into the ultraviolet.

These are not the only examples of nature using methods of optical engineering which have only recently become available to man. Butterflies of the Morpho group have a system of slats on the wing scales which act as multiple thin films and give rise to extremely beautiful iridescent colours. Squids and related animals have eyes which use multiple thin film mirrors as the focusing element rather than lenses. The colours of many of the iridescent feathers which make hummingbirds so attractive also arise from thin film effects. There is still much to learn from a study of the biological solutions to the problems of colour generation.

4.13 LOW EMISSIVITY WINDOWS

Windows in buildings are targets for improved energy efficiency. The reason for this is that normal window glass is an extremely good absorber and emitter of thermal energy. The black body equations given in Chapter 1 will show that a room with a temperature of 21°C has approximately 94 per cent of the thermal energy in the range 5–40 μm, with a peak at about 10 μm. Glass absorbs and re-emits about 80 per cent of this energy, making windows an appreciable gateway for loss of heat. Windows which address this problem are known as *low emissivity* windows. The effect is achieved by the use of multiple thin films.

The details depend upon the place of use. In colder regions it is not only necessary to minimise heat loss to the outside but also to guarantee that the solar energy penetrates the glass and acts as a passive heating agent. In desert regions it might be more desirable to reflect the external solar energy.

All the systems in use rely on coating the inside of the inner pane of a double sheet of glass with a thin film of strongly absorbing material, as in Figure 4.14a. One commonly used substance is tin dioxide, SnO_2, doped with fluoride ions, F^-. This material is able to strongly absorb the thermal energy from the room but has a low emissivity and hence cannot lose the energy by radiation. The energy is conducted back through the glass and returned into the room by radiation from the uncoated surfaces. The useful performance of the thin film is limited by its thickness. As films become thicker the emissivity increases, so it is important to keep film thickness low. Unfortunately, the ideal thickness for SnO_2 films is exactly that which produces a green colour due to interference and it has been found that customers do not like green-tinted windows. However, thin film effects can also provide a solution. The green reflection from the doped SnO_2 layer can

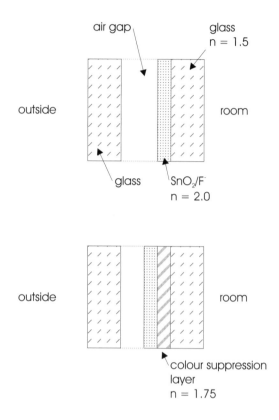

Figure 4.14 (a) Low emissivity coatings are applied to the inside of a double glass unit and on the side facing into the room . The thin films are often of tin dioxide, SnO_2, doped with fluoride ions, F^-. (b) Because the thin films strongly reflect in the green another thin film with a lower refractive index is applied between the SnO_2 and the glass as a colour suppression layer

be suppressed by coating the SnO_2 with a thin layer of transparent material with a lower refractive index than the SnO_2, as in Figure 4.14b.

4.14 ANSWERS TO INTRODUCTORY QUESTIONS

Why are soap bubbles coloured?
Soap bubbles are coloured because light is reflected from both sides of the film. The two resulting light waves interfere to produce constructive and destructive interference effects. The colours seen are a result of this process summed over all the visible wavelengths.

Why are thin films of oil on water coloured?

The reason is the same as in Q1. The colours will differ from those in soap films because of phase changes introduced at the oil–water interface which are not found at the water–air interface in the case of soap films.

How are antireflection coatings on lenses produced?

Antireflection coatings are produced by depositing a layer of a substance with a refractive index lower than the glass onto the lens surface. Usually a thickness of a quarter wavelength of yellow light is chosen. These films are not completely antireflecting because materials with a suitable refractive index are not available. For the best lenses multilayers can give better antireflection properties.

How can perfect mirrors be made from transparent materials?

Perfect mirrors can be made by building up multilayers of transparent materials on a glass substrate. Each layer is usually made a quarter of a wavelength thick and alternate materials in the layers have a refractive index higher or lower than that of the glass. Using computer programs, many sequences of layers can be devised which give virtually 100 per cent reflection.

4.15   FURTHER READING

Much of this chapter is concerned with thin film optical engineering. An introduction to the topic is given by:

O. S. Heavens and R. W. Ditchburn, Insight into Optics, Wiley, New York (1991), Ch. 4.

Complete coverage of the theory of single and multiple thin films is in:

H. A. McLeod. Thin Film Optical Filters, Hilger, London (1969).

The beautiful colours produced by thin films in nature are described and explained in:

C. Isenberg, The Science of Soap Films and Soap Bubbles, Tieto, Clevedon (1978).
D. K. Lynch and W. Livingston, Color and Light in Nature, Cambridge University Press, Cambridge (1995).

K. Nassau, The Physics and Chemistry of Color, Wiley-Interscience, New York (1983).

The colours of butterfly scales produced by microstructures are discussed by:

H. Ghiradella, Applied Optics, **30**, 1991, p. 3492.

The thin film mirrors in the eyes of squids are described by:

M. F. Land, Scientific American, **239**, December 1978, p. 88.

Free "Filmstar" software for the computation of thin-film optics is available from Dr F. T. Goldstein, FTG Software Associates, PO Box 597, Princeton, NJ 08542, USA, most easily obtained via: www.ftgsoftware.com/fsfree.htm.

4.16 PROBLEMS AND EXERCISES

1. Calculate the reflectivity of a plate of flint glass (n = 1.60) in air.

2. How will a film of optical thickness $\lambda/4$ in air appear in reflected light?

3. How will a film of optical thickness $\lambda/2$ in air appear in reflected light?

4. What is the minimum real thickness of a film of TiO_2 ($n = 2.90$) in air for constructive interference in yellow light with a wavelength of 550 nm? Estimate the colour that the film might appear when viewed in white light by reflection and in transmission.

5. Calculate the reflectivity of a thin film of optical thickness $\lambda/4$ and refractive index 1.5 in air.

6. Show that the reflectivity of a thin film of optical thickness $\lambda/2$ and refractive index 1.5 in air will be zero.

7. Compare the interference effects from films of optical thickness $\lambda/4$ and $\lambda/2$ in air with that of the same films deposited on a substrate with a higher refractive index.

8. Estimate the colour of a single thin film with a refractive index of 1.5, when viewed at 45° to the surface in air if the colour when viewed at normal incidence is third order green corresponding to a retardation, Δ, of 1350 nm.

THE COLOUR OF A THIN FILM IN WHITE LIGHT

9. Estimate the limiting colour of a thin film with a refractive index of 1.5, when viewed at grazing incidence to the surface if the colour when viewed at normal incidence is third order green corresponding to a retardation, Δ, of 1350 nm.

10. Write down expressions for the path difference that must hold for a thin film on a substrate in air to be highly reflecting (constructive interference) or not (destructive interference).

11. Derive the relationship between the refractive index of a thin film of refractive index n_f on a substrate of refractive index n_s which will act as (a) an antireflection coating or (b) a high reflectivity coating.

12. Determine the reflectivity of a MgF_2 film of $\lambda/4$ and refractive index 1.384 thickness on a crown glass lens of refractive index 1.523.

13. Calculate the reflectivity of a glass object of refractive index 1.52 coated with a $\lambda/4$ thickness film of TiO_2 with a refractive index of 2.9.

14. Determine the reflectivity in air of a $\lambda/4$ multilayer stack as a function of the number of pairs of low/high refractive index layers. Assume that the layers are deposited on a glass substrate, the low refractive index material is next to the substrate and the high refractive index material faces air. The refractive indices are: $n_0 = 1.0$(air), $n_s = 1.5$(glass), $n_H = 2.60$, $n_L = 1.384$.

APPENDIX 4.1   THE COLOUR OF A THIN FILM IN WHITE LIGHT

Retardation, nm	Colour reflected[a]	Colour transmitted[b]
0	Start of 1st order	Black, Bright white
40	Iron grey	White
97	Lavender-grey	Yellowish-white
158	Grey-blue	Brownish-white
218	Grey	Brownish-yellow
234	Green-white	Brown
259	White	Bright red
267	Yellow-white	Carmine red
281	Straw yellow	Deep violet
306	Bright yellow	Indigo
332	Yellow	Blue
Retardation, nm	Colour reflected[a]	Colour transmitted[b]

430	Yellow-brown	Grey-blue
505	Orange-red	Blue-green
536	Red	Green
551	Deep red	Yellow-green
555	end of 1st order	
	start of 2nd order	
565	Purple	Bright green
575	Violet	Green-yellow
589	Indigo	Gold
664	Sky blue	Orange
680	Blue	Orange-brown
728	Blue-green	Brown-orange
747	Green	Carmine red
826	Bright green	Purple-red
843	Yellow-green	Violet-purple
866	Green-yellow	Violet
910	Yellow	Indigo
948	Orange	Dark blue
998	Orange-red	Green-blue
1050	Violet-red	Yellow-green
1100	Dark violet-red	Green
1120	end of 2nd order	
	start of 3rd order	
1128	Blue-violet	Yellow-green
1151	Indigo	Off-yellow
1258	Blue-green	Pink
1334	Sea green	Brown-red
1350	Green	Purple-violet
1376	Dull green	Violet
1400	Yellow-green	Violet-grey
1426	Green-yellow	Grey-blue
1450	Yellow	Indigo
1495	Rose pink	Sea green
1534	Carmine red	Green
1621	Dull purple	Dull sea green
1650	Violet-grey	Yellow-green
1665	end of 3rd order	
	start of 4th order	
1682	Blue-grey	Green-yellow
1710	Dull sea green	Yellow-grey
1750	Blue-green	Lilac
1800	Green-brown	Purple-red
1811	Green	Carmine
1900	Pale green	Red
1927	Greenish-grey	Grey-red
2000	Pale grey	Blue-grey
2200	Very pale red-violet	Green-grey
2040	Red	Green
2240	end of 4th order	

Retardation, nm	Colour reflected[a]	Colour transmitted[b]
	start of 5th order	
~2500	Green	
~2700	Pink	
2800	end of 5th order	

[a] This colour is the same as that shown in *transmission* by a thin transparent plate of an anisotropic crystal viewed in white light between crossed polars.

[b] This colour is the complementary colour to that reflected. It is the same as that shown in *transmission* by a thin transparent plate of an anisotropic crystal viewed in white light between parallel polars. In addition these colours are seen in *reflection* when a thin transparent film on a substrate with a greater refractive index is viewed in white light.

CHAPTER 5
COLOUR DUE TO SCATTERING

Why is the sky blue and the sunset red?
Why are eyes blue at birth?
What produces the colour in a gold sol?

Scattering is a complex process that generally refers to the interaction of a light beam with small particles such as smoke, dust or water droplets. It is scattering which causes sunbeams to become visible in dusty or smoky rooms. Moreover, it is scattering in the atmosphere which produces the blue of the sky. In this chapter the way in which scattering can generate intense colours will be presented.

5.1 SCATTERING

For many purposes it adequate to define scattering in an observational way. If a transparent medium contains scattering centres the intensity of light traversing the medium in the incident direction will gradually fall as the light is scattered into other directions. The reduction in the intensity of a beam of light which has traversed a medium containing scattering centres can be written as

$$I = I_0 \exp(-\alpha_s l) \tag{5.1}$$

where I_0 is the incident beam intensity, I the intensity after travelling a distance l in the turbid medium and α_s is an experimentally determined linear *scattering coefficient*. The form of this equation is identical to that of Lambert's law for absorption, but now scattering centres have been substituted for absorbing centres. The scattering coefficient is found to depend upon:

- the number of scattering centres present
- the ratio of the particle diameter to the wavelength of the light

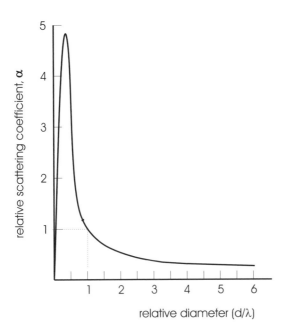

Figure 5.1 Schematic illustration of the effect of particle size, expressed as relative diameter (particle diameter, d / wavelength of light, λ) on the relative scattering coefficient, α, expressed as the ratio of the scattering coefficient α_s at the relative diameter of 1.0 to that at other relative diameters. The maximum scattering occurs when the particle diameter is about half the wavelength of the light

- the ratio of the refractive indices of the particle and the surrounding medium
- the particle shape (see Section 5.8 for an example of this effect).

If all these parameters are held constant except for particle size, maximum scattering is found to occur when the particle size is somewhat less than the wavelength of the light illuminating the system. This is illustrated schematically in Figure 5.1.

The effects just described are well known in industry. For example, many plastics are made opaque by the addition of white pigment. Most frequently this is titanium dioxide, TiO_2, but china clay and limestone are also commonly employed. It is known that the scattering power of the pigment particles depends upon particle size and that the maximum opacity of the plastic occurs when the diameter of the particles is about half the wavelength of light, that is, about 200 nm.

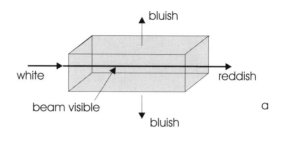

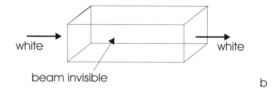

Figure 5.2 (a) Small particles suspended in a liquid will preferentially scatter blue light. The transmitted light will take on a reddish colour and the beam will be visible in the liquid. (b) In a true solution no scattering takes place and the beam remains invisible

While this phenomenological approach is adequate for many purposes it is not detailed enough for an understanding of how scattering can produce colours. This topic occupies the following sections.

5.2 TYNDALL BLUE AND RAYLEIGH SCATTERING

The scattering of light by small particles was studied by a number of scientists in the 19th century, but the most detailed experiments were made by Tyndall. He observed that liquids containing suspensions of small droplets, such as water containing a little milk, looked sky blue when illuminated with white light and viewed from the side. The beam of light responsible was also visible in the liquid and the light emerging in the beam direction took on a red hue. This is shown schematically in Figure 5.2a. The fact that a beam of light is visible in a suspension of small particles but invisible in a true solution (Figure 5.2b) is still the easiest way of distinguishing one from the other. Tyndall supposed (correctly) that blue light was scattered more strongly than red light, and this blue scattering is still referred to as *Tyndall blue*.

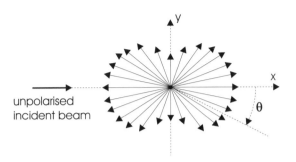

Figure 5.3 The Rayleigh scattering pattern of unpolarised light from small particles. The lengths of the arrows diverging from the small scattering centre can each be thought of as defining the scattered intensity at a distance r and at an angle θ to the forward direction

The mathematics of scattering were first derived by Rayleigh, who showed that even fluctuations in the refractive index of an otherwise homogeneous material could cause scattering. The theory is expressed in its simplest form for spherical insulating particles with a diameter less than one tenth of the wavelength of the incident light. Scattering by such bodies is referred to as *Rayleigh scattering*. In the case where a beam of unpolarised light of intensity I_0 is scattered once only by the scattering centre, the intensity of the scattered light, I_s is given by:

$$I_s = I_o \left(9\pi^2 V^2 / 2r^2\lambda^4\right) \left[(m^2 - 1) / (m^2 + 2)\right]^2 (1 + \cos^2\theta) \qquad (5.2)$$

where the measurement is taken at a distance r from the scattering centre, V is the volume of the scattering particle, λ is the wavelength of the light and θ is the angle between the incident beam and the direction of the scattered beam as shown in Figure 5.3. The quantity m is the relative refractive index of the particle:

$$m = n_{particle} / n_{medium}$$

In this case $n_{particle}$ is the refractive index of the particle and n_{medium} the refractive index of the surrounding medium. For air n_{medium} is 1.0.

If the intensity of the scattered light in a plane containing the incident beam, the scattering volume and the observer, the *plane of observation*, is plotted, a characteristic *Rayleigh scattering pattern* is formed, as in Figure 5.3. It indicates that as much light is scattered backwards as forwards and that only half as much intensity is scattered normal to the beam direction.

As the formula shows, all wavelengths scatter in this pattern, but the shorter wavelengths are more strongly scattered than the longer wavelengths.

The importance of the equation went beyond simply explaining scattering. A comparison of measurements of scattering with the theory made it clear that *molecules alone* could operate as scattering centres. That is, even the purest gas would still show light scattering. Moreover, the formula allowed an estimate of molecular size and the number of molecules present in a unit volume of a gas to be made. These values permitted scientists to estimate Avogadro's number and the molar masses of gases. Such information was of great interest towards the end of the 19th century, when the atomic theory of matter was still a topic of controversy.

5.3 BLUE SKIES, RED SUNSETS

The blue colour of the sky has been a topic of interest since antiquity. Newton made the reasonable suggestion that it arose by reflection from small water droplets in the atmosphere. Rayleigh showed that it was due to scattering by gas molecules in the atmosphere.

Because the scattering of light is proportional to $1/\lambda^4$, violet light is scattered far more than red light, as can be judged from Figure 5.4. This suggests that when we look at the sky in a direction which is not towards the sun, the colour seen should be indigo or violet. In fact the sky appears to be blue. This is for two reasons. First, the solar energy reaching the ground has less intensity in the violet than at longer wavelengths such as yellow. Second, the sensitivity of the eye to colour peaks in the yellow-green region of the spectrum, near to 555 nm (see Figure 1.5). The result of these factors is that the sky is perceived to be blue, as illustrated schematically in Figure 5.5a. Towards sunset, when it is possible to look in the direction of the sun through a thicker layer of atmosphere, the scattering will remove blue light preferentially and the sun and sky will appear red, as shown in Figure 5.5b. The effect will be enhanced when this light is reflected at a shallow angle from clouds or fine dust in the upper atmosphere, as can occur after volcanic eruptions, when spectacular sunsets are often recorded.

5.4 SCATTERING AND POLARISATION

A characteristic of the Rayleigh scattering curve drawn in Figure 5.3 is that it produces strongly polarised light. This can be illustrated by reference to

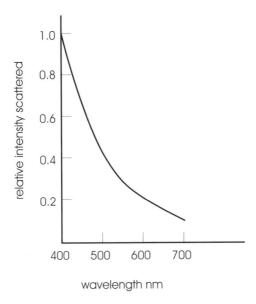

Figure 5.4. Rayleigh scattering of visible light as a function of wavelength. This is proportional to $1/\lambda^4$, and so violet light is scattered about eight to ten times more than red light

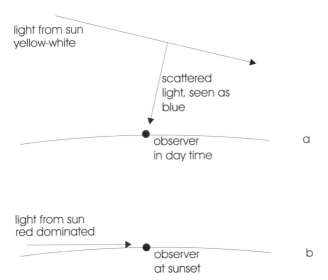

Figure 5.5 (a) An observer during the day will see light from the sun as yellow white, and scattered light, in other directions, will give the sky a blue colour. (b) At sunset, light in the direction of the sun will be depleted in blue and appear red, while the sky overhead will remain blue due to the scattered light

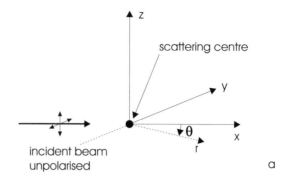

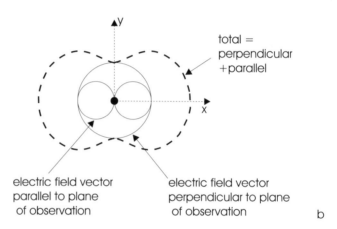

Figure 5.6 (a) A beam of unpolarised light travelling along the positive x-direction can be resolved into two linearly polarised components parallel and perpendicular to the x–y plane. This is taken as the plane of observation of the scattered light, which is observed at a distance r and angle θ to the positive x-direction. (b) The Rayleigh scattering pattern is made up of the sum of light scattered with its electric field vector perpendicular and parallel to the plane of observation

Figure 5.6a. Assume that an unpolarised light beam is travelling along the positive x-direction and that observations of the scattered light, (at a distance r and angle θ to the positive x-axis) are made in the x–y plane, which thus becomes the plane of observation. Any incident unpolarised light beam passing along the x-axis can be resolved into two linearly polarised components, one with the electric field vector lying parallel to the x–y plane and one with the electric field vector lying perpendicular to the x–y plane. The wave polarised perpendicular to the plane of observation

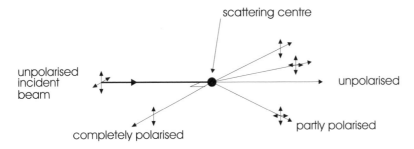

Figure 5.7 The polarisation of scattered light is a maximum in the direction perpendicular to the direction of the incident beam. In other directions in the plane of observation it is partially polarised

is found to be scattered equally in all directions, and is drawn as a circle in Figure 5.6b. This wave contributes the factor 1 in the $(1 + \cos^2\theta)$ term of Equation 5.2. The wave with the electric field vector in the plane of observation has a dumbbell intensity shape, shown in Figure 5.6b. This arises from the $\cos^2\theta$ factor in the $(1 + \cos^2\theta)$ term in Equation 5.2. The total scattering curve is the sum of both of these contributions.

When the scattered light is viewed at 90° to the incident direction only one polarisation component will be present and the light will appear to be completely linearly polarised, as drawn in Figure 5.7. In other directions a mixture of both polarisations is present. In the forward (0°) and backward (180°) directions these are equal to each other and the light is unpolarised.

Sky light is polarised due to this differential scattering. The degree of polarisation is least (virtually zero) in the direction of the sun. However, sky light in a plane which includes the observer and is at 90° to the line joining the observer to the sun, is strongly polarised as indicated in Figure 5.8. In fact the light should be completely polarised, as stated above, but in reality it is found to be only about 75–85 per cent polarised. In other directions the degree of polarisation lies between zero and this value. The reason for the discrepancy is that the actual polarisation observed at any point in the sky is a result of multiple scattering, the atmospheric conditions and the relative positions of the sun and the observer. In fact the accurate evaluation of the polarisation of the sky light is complex, and it is only in the second half of the 20th century that accurate polarisation maps of the sky have been produced with the aid of high speed computers.

The polarisation of the sky can easily be observed using uniaxial or biaxial crystals. Cordierite (a magnesium aluminosilicate, $Mg_2Al_4Si_5O_{18}$, with the beryl structure) is a biaxial mineral and absorption of polarised

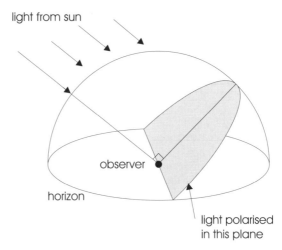

light from sun

observer

horizon

light polarised
in this plane

Figure 5.8 Light scattered from small molecules in the air is optimally polarised in a plane at 90° to the direction of the incident radiation. It is not completely polarised in this plane due to multiple scattering. Observation of the polarisation of the light of the sky will allow the observer to estimate the position of the sun even on overcast days

light is strong along only one crystallographic axis. It is recounted that Vikings used this property of cordierite crystals, called "sunstones", to locate the sun (and so navigate) even on cloudy days. The sky is viewed through a cordierite crystal which is rotated at the same time. If the direction of observation is in a plane perpendicular to the direction of the sun the sky will appear alternately darkened and brightened as the crystal rotates. Viewing in the direction of the sun does not produce this effect as the light is not very polarised. Hence the direction of the sun can be determined even on overcast days. The effect is easily checked with a piece of Polaroid film.

Humans are unable to detect polarised light[1] but bees and ants, and perhaps many other insects possess this ability. They use this skill to navigate to and from the hive or nest even under conditions when the sun is hidden from them. The capability arises because the molecule responsible for photoreception in the eyes of all animals, rhodopsin (see Chapter 8), is a dipolar molecule with an optical axis. These molecules absorb polarised light energy maximally when the direction of polarisation is parallel to the

---

[1] In fact this is not altogether correct. The visual phenomenon called Heidinger's brushes, a faint small hourglass shape centred in the field of view, is caused by the detection of polarised light in the retina of the eye. It is not observed by everyone.

optical axis of the molecule. In insects' eyes these molecules are aligned in a fixed direction, making them polarisation sensitive. In humans the molecules are free to rotate, so that the orientation of the optical axis is random and polarisation perception is lost.

5.5 BLUE EYES AND SOME BLUE FEATHERS

Most eyes are blue at birth. This is, in fact, a scattering effect. The iris of the eye is a composite of various tissues, small crystalline regions and air vesicles, each of which have differing refractive indices. This inhomogeneity gives rise to preferential scattering of blue light. As the light transmitted through the lens and iris is absorbed in the underlying tissues, only the scattered light re-emerges, to give the impression of blue irises. Many irises develop a pigment a few days after birth and it is this pigment that changes the colour from blue to green or brown.

The vivid blue colours of many feathers from exotic birds are coloured blue in the same way. The outer part of the feather is a composite of several different proteins together with small air vesicles. The inhomogeneity causes preferential blue scattering. This is not easily seen against a background of reflected light, but if the feathers are backed by a dark absorbent layer, the blue colour becomes easily visible.

5.6 MIE SCATTERING

The mathematics of the problem of scattering is formidable and the first complete theory for spherical particles was only completed in 1908 by G. Mie. Once again, though, each photon was presumed to be scattered only once. The term Mie scattering is generally reserved for scattering by particles which are somewhat larger than those for which Rayleigh scattering is valid, say about one third the wavelength of light or more. The scattering curve for Rayleigh scattering, illustrated in Figures 5.3 and 5.9a, gradually alters as the particle size increases. It is found that forward scattering begins to dominate over backward scattering, as in Figure 5.9b. As particle size passes the wavelength of light, the forward scattering lobes increase further still and side bands develop, representing maxima and minima of scattering at definite angles, as in Figure 5.9c. The position of these lobes depends upon the wavelength of the scattered light and so they are strongly coloured. These coloured bands, referred to as *higher-order*

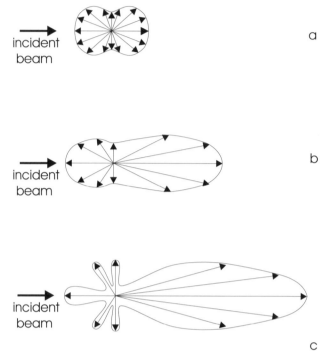

Figure 5.9 Schematic illustration of the patterns of intensity scattered by small particles. (a) Rayleigh scattering from particles much smaller than the wavelength of light. (b) For particles approaching the wavelength of light the scattering becomes pronounced in the forward direction. (c) For particles larger than the wavelength of light lobes of intensity appear which are wavelength dependent and so give rise to colours at specific viewing angles. These are called higher order Tyndall spectra

Tyndall spectra, are dependent upon the particle size and so can be used for particle size determination.

With even larger particles, white light becomes reflected (rather than scattered as we are discussing here) evenly in all directions. This is the situation which holds in fogs and mists.

5.7 GOLD SOLS AND RUBY GLASS

Gold sols, first prepared by Faraday utilising the chemical reduction of gold chloride solution, are brightly coloured. The colour is due to white light absorption and scattering from microscopic crystallites of gold which are small enough to remain suspended in aqueous solutions.

Ruby coloured glass has been known for much longer, having been produced since the 15th century. The process involved gold as the colorant. Modifications in the processing also allowed craftsmen to make blue or purple glass. Ruby glass is made by dissolving of the order of 0.01% of gold in molten glass. If the glass is cooled in a normal way, which is fairly rapidly, the glass remains clear as the gold atoms are distributed evenly throughout the material. Colour is developed by annealing (reheating) the glass to 650°C for several hours. At this temperature the gold atoms in the glass aggregate to produce gold crystals with a diameter of between 40 nm and 140 nm, distributed throughout the glass matrix. The colour is caused by these crystallites.

A precise explanation of the colour had to wait until 1908, and the theory derived by Mie for this purpose. In the previous discussions we have assumed that the particles which are scattering light are insulators and do not absorb light. However, metal particles are strongly absorbing. For strongly absorbing systems it is found that although the amount of light *scattered* is still proportional to $V^2\lambda^{-4}$ (the Rayleigh dependence) the *absorption* of light is proportional to $V\lambda^{-1}$, where V is the volume of the particles which are interacting with the incident light. Now as V becomes smaller the main interaction with light changes from scattering to absorption. This is seen in the curves reproduced in Figure 5.10. In the case of gold crystals with a diameter of 40 nm, Figure 5.10a, the absorption is dominant and the total light removed by absorption plus scattering peaks at 530 nm, in the green region of the spectrum. The transmitted colour lacks this wavelength and imparts a ruby red colour to the glass. (The red colour of ruby gemstones and the synthetic ruby crystals used in the first lasers arises for a different reason, and is treated in Chapter 7.) For particles of diameter 100 nm, Figure 5.10b, the scattering is now larger (and comparable to) the absorption and the total light removed peaks at 580 nm. This imparts a purple transmission colour to the glass. At even larger crystallite sizes of 140 nm, Figure 5.10c, the scattering is now much larger than the absorption at the red end of the spectrum and the total of the light absorbed and scattered peaks in the red at 610 nm. The glass now looks blue in transmission. Further increase in size leads to the domination of scattering, and ultimately reflection, over absorption.

Other colours can be produced in glass by using other noble metals, notably silver for yellow and platinum for pink. Control of the crystallite size and hence colour by processing is difficult, which is why early glassworkers who had perfected recipes for the production of ruby glass guarded their knowledge jealously.

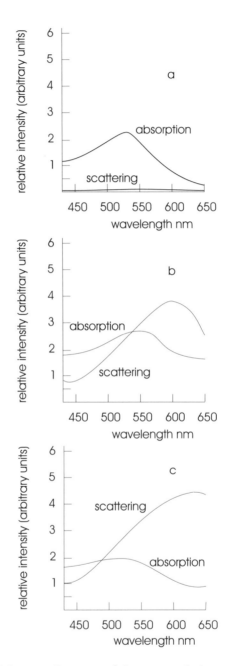

Figure 5.10 Schematic illustration of the way in which scattering and absorption change with radius for spherical gold particles: (a) diameter 40 nm; (b) diameter 100 nm; (c) diameter 140 nm. The intensity scale is in arbitrary units

5.8 POLYCHROMIC GLASS

It has been possible to make beautiful *polychromic* glass with almost any spectral colour using the absorption characteristics of metal particles. The processing however, is more complex than in the case of ruby glass. The method is rather similar to that used to prepare photochromic sunglasses, which darken on exposure to sunlight. An outline follows.

1. A glass composition is made up with additives of silver bromide, AgBr, cerium dioxide, CeO_2, and sodium fluoride, NaF. These additives dissolve in the glass and on cooling the glass blank remains clear. Glasses are usually made in furnaces under slightly reducing conditions and the cerium present is found as ions in the Ce^{3+} state.
2. The glass is exposed to ultraviolet light for about 1 min. This energy is absorbed by the Ce^{3+} ions and electrons are liberated via the process:

$$Ce^{3+} + h\nu \text{ (ultraviolet)} \rightarrow Ce^{4+} + e^-$$

These electrons are taken up by silver ions to form silver atoms.

$$Ag^+ + e^- \rightarrow Ag$$

3. A heat treatment at about 450°C allows the silver atoms to diffuse towards each other and so form small crystallites distributed throughout the glass.
4. Further heat treatment at about 500°C causes the halide additives to crystallise on the silver nuclei in the following unusual way. Sodium fluoride first grows on the silver as small cubes. Pyramids of a mixed sodium silver bromide phase, (Na,Ag)Br, then grow on the cubes of NaF, as illustrated in Figure 5.11. The glass remains colourless when the crystals are below about 200 nm in size as they are too small to scatter light appreciably. (If the crystallites become much larger than this they scatter light and the glass becomes hazy or opalescent and has to be rejected.)
5. The clear glass is again exposed to ultraviolet light, this time for periods of up to 2 h. This causes further precipitation of silver atoms in the tips of the pyramidal crystallites, by the same mechanism as above.
6. Another annealing at about 350°C causes the Ag atom clusters to grow as needle-like crystals an the tips of the (Na,Ag)Br pyramids, as depicted in Figure 5.11.

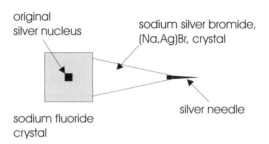

Figure 5.11 The form of crystallites in polychromic glass. The silver needles at the tip of the sodium silver bromide, (Na,Ag)Br, pyramids cause intense colours by selective light absorption

Table 5.1 Colour and needle dimensions in polychromic glass

Needle length, nm	Needle width, nm	Length/width	Colour transmitted
3.0	3.0	1.0	Yellow
4.0	3.0	1.3	Deep yellow
5.0	3.0	1.7	Orange
6.0	3.0	2.0	Orange-red
7.5	3.0	2.5	Red
10.0	3.5	2.9	Magenta
12.0	3.5	3.4	Purple
16.0	4.0	4.0	Blue
21.0	4.5	4.7	Turquoise
36.0	6.0	6.0	Green

The process is designed to form needles with dimensions of between 3 nm and 6 nm wide and 3 nm and 36 nm in length. They are too small to cause much light scattering, but do absorb strongly, and generate brilliant colours when the glass is viewed in transmission. The precise absorption characteristics depend critically on the needle shape, especially the ratio of the width to the length. Some information on the relationship between particle geometry and colour transmitted is given in Table 5.1.

Careful control of processing allows manufacturers to produce a full range of colours throughout the body of the glass.

5.9 SCATTERING AND TRANSPARENCY OF GLASS

Opalescence, which causes glass to be translucent rather than transparent, is useful in light diffusers and for decorative purposes. The effect is brought about in clear glass by the introduction of white particles, often TiO_2, as remarked at the beginning of this chapter. However, the primary quality needed for most optical purposes is high transparency. Glasses for this use are therefore made carefully so as to eliminate all scattering centres, such as voids or impurity particles, as these reduce the transparency considerably. It comes as a surprise to find that certain partly crystallised glasses have been found with a higher transparency than the pure glass. How this comes about tells us something about the complexity of the scattering process.

In experiments to make glasses containing lanthanide ions for up-conversion purposes (see Chapter 9) some slightly crystallised glasses were found to be *more* transparent than the original glass. The glasses were made from sodium oxide, aluminium oxide and silica (Na_2O-Al_2O_3-SiO_2) together with small amounts of lanthanum trifluoride, LaF_3. The initial glass was usually clear. Heating this for several hours at temperatures in the range 800–850°C caused crystallisation of LaF_3 to occur and a glass ceramic[2] formed. As expected, the materials were opaque, due to the formation of large crystals in the matrix. However, it was surprising to find that glass heated at lower temperatures for short times were sometimes more transparent than the parent glass. Some data linking crystallite size in the glass and heat treatment are given in Table 5.2.

The transparency of the glass can be expressed by Lambert's law:

$$I = I_0 \exp(-\alpha l)$$

where I_0 is the incident beam intensity, I the intensity after travelling a distance l in the glass medium and α is the total linear absorption coefficient, measured in m^{-1}, which takes into account loss of intensity from both scattering centres and absorption centres. Reference to Table 5.2 shows clearly that the glasses heated at 750 and 775°C both have lower absorption coefficients than the starting material, even though they contain small crystallites.

How can this counter-intuitive result be explained? The reason is that in the formulae presented earlier in the chapter, it was assumed that the scattering centres were rather far apart and that each photon was only scattered once. This is not true in the present system and multiple

---

[2] Glasses which have been deliberately recrystallised produce solids called *glass ceramics*. They are composed of a mass of interlocking crystallites in a matrix consisting of the remaining glass. Although these materials have lost transparency, other properties, particularly mechanical strength and thermal shock resistance, are greatly enhanced. This has made them widely used in day to day applications such as cooking ware.

Table 5.2 The increase in transparency of glass containing LaF_3 crystallites[a]

Heating temperature (°C)[b]	Crystal size (nm)	Absorption coefficient (cm^{-1}) at 620 nm
–	–	0.075
750	7.2	0.060
775	12.4	0.070
800	19.6	0.095
825	33.3	0.101
850	–	Opaque

[a] These data are extracted from information given by M. J. Dejneka, MRS Bulletin, **23**(November), 1998, p. 57.
[b] Heating time 4 h for each sample.

scattering due to closely spaced scattering centres is the normal state of affairs. Computer calculations of multiple scattering effects in this system have been used to verify that the observed result is indeed possible.

5.10 Answers to Introductory Questions

Why is the sky blue and the sunset red?
Strictly speaking the sky is blue when observed in a direction away from the direction of the sun. The blue colour occurs because molecules in the atmosphere scatter blue light, by Rayleigh scattering, much more than red light. Sunsets are red because the light coming directly from the sun is diminished in blue, by the same Rayleigh scattering mechanism, so as to give it an overall red colour. This is made easier to see when the light is reflected from clouds or dust in the upper atmosphere.

Why are eyes blue at birth?
The colour of eyes resides in the iris. At birth this is usually unpigmented and the colour is caused by scattering from the inhomogeneous structures in the iris. As blue light is scattered preferentially and all non-scattered light is absorbed behind the iris the colour can look intense.

What produces the colour in a gold sol?
The colour in a gold sol is seen in transmission. It is produced by small crystals of gold which have been deliberately precipitated in the solution during processing. The size of the crystallites is closely controlled so as to give a colour due to a mixture of absorption and scattering.

5.11 FURTHER READING

The colours produced by scattering in the atmosphere are explained in:

D. K. Lynch and W. Livingston, Color and Light in Nature, Cambridge University Press, Cambridge (1995), Ch. 2.

J. Walker, Scientific American, **260**, January 1989, p. 84 and references therein.

The mathematics of atmospheric scattering are treated in detail by:

E. J. McCartney, Optics of the Atmosphere, Wiley, New York (1976).

The use of cordierite crystals for navigation, the navigation of insects using polarised light and information about the polarisation of sky light is given in:

A. Nussbaum and R. A. Phillips, Contemporary Optics for Scientists and Engineers, Prentice-Hall, Englewood Cliffs, NJ (1976), p. 369.

R. Wehner, Scientific American, **235**, July 1976, p. 106.

J. Walker, Scientific American, **238**, January 1978, p. 132.

The theory of Mie scattering and the colours of gold and other sols is given by:

H. C. van de Hulse, Light Scattering by Small Particles, Wiley, New York (1957), Ch. 19.

Polychromic glass is described in:

D. M. Trotter, Scientific American, **264**, April 1991, p. 56.

S. D. Stookey, G. H. Beall and J. E. Pierson, J. Applied Physics, **49**, 1978, p. 5114.

The discovery of enhanced transmission in LaF_3 glass ceramics is reported by:

M. J. Dejneka, MRS Bulletin, **23**(November), 1998, p. 57.

5.12 PROBLEMS AND EXERCISES

1. Estimate the amount of Rayleigh scattering for wavelengths 450 nm, 500 nm, 550 nm, 600 nm, 650 nm and 700 nm relative to that of violet (400 nm) and compare your results with Figure 5.4.

2. The sky looks blue and yet when sunlight is passed through a prism (Newton's experiment) all colours of the spectrum are seen. Why is the sky not seen as white?

3. The blue colour of eyes and most biological blue colours are greatly enhanced by a backing of dark material. Why does the sky appear so intensely blue?

4. Assuming that no other factors besides particle size are involved in the discussion of gold sols, at what particle size does the scattering become as important as absorption?

5. Using the data contained in Table 5.1, make a qualitative estimate of the colour you would expect to be transmitted by a sample of polychromatic glass containing needles with a length / width ratio greater than 6.0?

CHAPTER 6

COLOUR DUE TO DIFFRACTION

How can you use a steel ruler or a hi-fi record to measure
the wavelength of light?
Why are opals coloured?
How do liquid crystal thermometers work?

Diffraction effects occur when waves interact with objects having a size
similar to the wavelength of the radiation. Although often observed with
water waves, diffraction of light is not commonly encountered in everyday
life. Nevertheless it is important and it is diffraction that often sets a limit
to the performance of optical instruments.

Diffraction is a complex process and its explanation is mathematically
involved. In general, two regimes have been explored in most detail: (i)
diffraction quite close to the object which interacts with the light, called
Fresnel diffraction; and (ii) the effects of diffraction far from the object
which interacts with the light, called *Fraunhofer diffraction*. The result of
diffraction is a set of bright and dark fringes, due to constructive and
destructive interference, called a *diffraction pattern*. When white light is
involved intense colours can be observed.

6.1 DIFFRACTION AND COLOUR PRODUCTION BY AN APERTURE

If a long narrow slit is illuminated by monochromatic light the intensity
pattern observed far from the slit (the *Fraunhofer diffraction pattern*) is
given by the expression:

$$I_x = I_0 \left(\frac{\sin x}{x} \right)^2$$

where $x = \pi w \sin \theta / \lambda$ and w is the width of the slit, θ is the angular deviation
from the "straight through" position, and λ is the wavelength of the light.
This produces a set of bright and dark fringes with *minima* given by:

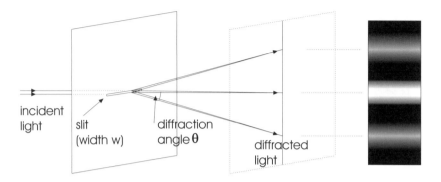

Figure 6.1 The fringes produced by diffraction of monochromatic light by a long narrow slit of width w. Diffracted light is concentrated into bands at various values of the angle $\pm\theta$ to the undeviated beam. The minima between the bright bands is given by sin θ = 1/w, 2/w, 3/w, etc.

$$\sin \theta_{min} = m\lambda / w$$

where m takes values $\pm1\pm2\pm3$, etc. For θ_{min} to be appreciable, w must be close to λ. In fact the formula shows that the spacing between the minima will be proportional to the reciprocal of the slit width, so that the narrower the opening the wider will the fringe spacing be. The effect is illustrated in Figure 6.1. The positions of the *maxima* between these dark bands are not given by such a simple formula, but can be *approximated* by assuming that they lie midway between the minima.

It is seen that the sine of the angle through which a ray is diffracted is related to its wavelength. This indicates that each wavelength in white light will be diffracted through a slightly different angle and that red light will be diffracted through a greater angle than violet light. In this way white light will produce a set of diffraction patterns, each belonging to a different wavelength, as sketched in Figure 6.2. These patterns look like, and are called, spectra. They are referred to as *first order*, *second order* and so on as they are recorded further and further from the undeviated beam.

When the slit is shortened so as to form a rectangular aperture the diffraction maxima will take the form of small rectangular spots running in two perpendicular directions, as depicted in Figure 6.3a,b. White light will produce coloured spots via the same mechanism as described above.

The form of the diffraction pattern produced by a circular aperture, Figure 6.3c, can be inferred by reference to the rectangular aperture. The diffraction pattern will consist of a series of bright and dark circles

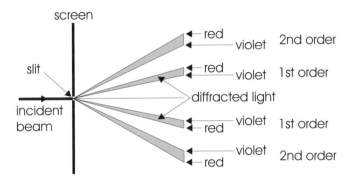

Figure 6.2 The diffraction patterns, resembling spectra, produced by diffraction of white light by a long narrow slit. The first-order patterns are those closest to the undeviated beam. The angle through which red light is deviated is greater than that by which violet light is deviated for each order of diffraction

concentric with the original aperture. The spacing of the maxima and minima is given by:

$$\sin \theta = n\lambda / d$$

where θ is the angle between the directly transmitted ray and the diffraction ring as indicated in Figure 6.3d, λ is the wavelength of the light and d the diameter of the aperture. The computation of n requires rather sophisticated mathematics, the results of which show that n takes the values 0 (central bright spot), 1.220 (first dark ring), 1.635 (first bright ring), 2.333 (second dark ring), 2.679 (second bright ring), 3.238 (third dark ring).

Just as with the slit, the dependence of the diffraction angle upon wavelength means that a circular aperture illuminated with white light will produce a set of coloured rings, rather like miniature circular rainbows. The formula indicates that each ring will have a violet inner edge and a red outer edge. For a similar reason a multicoloured ring can sometimes be seen to surround a narrow beam of white light which has passed through a pane of glass covered with a fine powder or with fine drops of moisture. Each particle diffracts as a small circular aperture. The eye intercepts many of these diffracted rays and a coloured ring is seen which is composed of fragments of colour from many different dust particles. As before, red is on the outside of the circle and violet on the inside. The same effect can sometimes be seen around the image of a small light in a dusty mirror. Again each dust particle acts so as to diffract the light, which is reflected from the mirror surface back towards the observer.

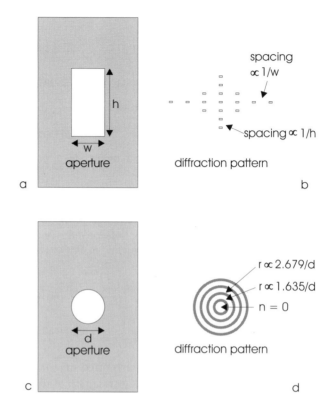

Figure 6.3 Diffraction fringes from a rectangular aperture (a) consist of perpendicular rows of rectangular spots with spacings inversely proportional to the width and height of the aperture (b). The diffraction pattern from a circular aperture (c) consists of a bright central disc (Airy's disc) surrounded by a set of circular light and dark rings (Airy's rings) (d). The radii of the rings are inversely proportional to the aperture diameter

Diffraction by circular apertures plays an important part in the overall performance of many optical instruments such as telescopes and microscopes. The resolution of such instruments, which is, roughly speaking, equivalent to the separation of two points which can just be distinguished as separate objects, is controlled by diffraction. It is of the order of the wavelength of the observing radiation. Because of this limitation optical microscopes are unable to image atoms. Electron microscopes, using radiation with a wavelength of the order of 0.002 nm, are able to do so.

The performance of a telescope can be estimated by the appearance of the image of a star. Under ideal conditions the image will appear as a small point-like disc of light surrounded by diffraction rings. These are much

fainter than the disc and can often be more easily distinguished if the
eyepiece is pulled in or out by a short distance so as to defocus the image
slightly. The appearance of these images was first interpreted by Airy and
they are known as Airy's disc (for the central region) and Airy's rings (for
the surroundings). A distortion of Airy's rings, when atmospheric conditions are good, is indicative of poor optics. Exactly the same effect will be
seen if a point of light is observed in an optical microscope. Slight
defocusing will reveal an expanding set of Airy's rings, the perfection of
which mirrors the perfection of the lenses.

6.2 COLOUR PRODUCTION BY DIFFRACTION GRATINGS

Planar diffraction gratings consist of an object inscribed with a set of
parallel lines which have a spacing similar to that of the wavelength of light.
A *transmission grating* has alternating clear and opaque lines and diffraction effects are observed in light which has been allowed to pass through
the clear strips. A reflection grating consists of a set of grooves or blazes
and diffraction effects are observed in the light reflected from the patterned
surface. The effectiveness of a grating is the same whether light is transmitted through it or reflected from it.

The positions of the diffraction *maxima* for a *transmission grating*
when illuminated by monochromatic light normal to the surface is given by
the formula:

$$\sin \theta = n\lambda / d$$

where d is the repeat spacing of the grating, λ is the wavelength of the
radiation and θ is the angle through which the beam in question has been
diffracted as depicted in Figure 6.4a.

When light falls on a *reflection grating* at close to grazing incidence, as
in Figure 6.4b, the formula for the positions of the diffraction *maxima* is:

$$(1 - \cos \theta) = n\lambda / d$$

where d is the repeat spacing of the grating, λ is the wavelength of the
radiation and θ is the angle through which the beam in question has been
diffracted as depicted in Figure 6.4b.

The term n in both formulae can take integer values of 0, ±1, ±2 and so
on. Each of these corresponds to a different diffraction maximum, called an

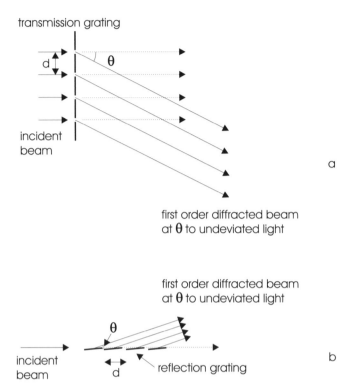

Figure 6.4 (a) Diffraction by a transmission diffraction grating. The first-order pattern lies at an angle ±*θ* to the undeviated beam given by sin *θ* = λ/d. (b) Diffraction by a reflection diffraction grating. The first-order pattern lies at an angle *θ* to the undeviated beam given by (1 − cos *θ*) = λ/d

order. When illuminated by white light, each wavelength will be diffracted through a slightly different angle so that each order will consist of a spectrum. These are similar to those produced by a long narrow slit, shown in Figure 6.2, but because each line on the grating acts as a contributing slit they are of much greater intensity. Diffraction gratings can thus give rise to very intense colours. These are very easily seen when the image of a small white light is viewed by reflection from the surface of a gramophone record. Plastic transmission gratings are inexpensive and can be used to show the spectra of many light sources, including those from street lights, as is remarked in Chapter 7.

Diffraction of light from two-dimensional gratings, which produce a rectangular array of diffraction spots similar to those from a single rectangle (Figure 6.3), is easily made visible. If a far off sodium street lamp is viewed through fairly closely woven net curtains, perpendicular rows of yellow spots will be seen centred upon the image of the light itself. When white light is observed, coloured patterns can be detected. The intensity of the patterns is low and are seen to best advantage when viewing a small distant bright light through a closely woven black fabric such as an opened umbrella, which absorbs superfluous reflection and scattering.

6.3 ESTIMATION OF THE WAVELENGTH OF LIGHT BY DIFFRACTION

It is surprisingly easy to estimate the wavelength of light using a hi-fi gramophone (phonograph) record or even a steel ruler in conjunction with the phenomenon of diffraction. The method uses the rule or record as a reflection grating. Set up a pen torch or similar small light and observe its image reflected from your grating. A steel rule graduated in 1/2 mm can be used but a hi-fi gramophone record is better because it has a closer spacing. (It is also possible to buy inexpensive plastic diffraction gratings with about 1000 lines per mm. These are even better and spread the spectra very widely.) As the grating is tilted several coloured orders of diffraction will be seen to either side of the central reflection of the torch light. The spacing of the spectra will depend upon the angle of incidence and the angle of reflection, as well as the distance to the light and will change as the grating is rotated or as the light is approached.

To measure the wavelength of light without measuring the angles of incidence and reflection it is necessary to view the spectra at grazing incidence. Mount a piece of white paper clearly marked in centimetres, so as to form a scale running in a vertical direction below the light. Arrange the grating horizontally and observe the first order spectrum by raising your eyes just above the surface of the grating. With a little practice it is possible to use both eyes so as to see the coloured pattern reflected from the grating superimposed upon the vertical scale. The arrangement is drawn Figure 6.5a. Note the distance, d, on the scale against which any chosen colour appears and then measure the distance, D, of the lamp from the grating. The details of the geometry are given in Figure 6.5b. In order to see a diffraction pattern the path difference between light reflected from adjacent grooves or marks must be an integral number of wavelengths of the appropriate light. Reference to Figure 6.5b shows this to be:

$$n\lambda = AC - BC$$

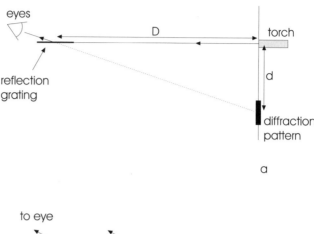

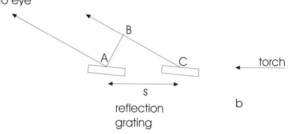

Figure 6.5 (a) The arrangement to measure the wavelength of light with a simple reflection grating such as a steel rule or a hi-fi record. The distance of the grating from the torch is D. Using both eyes the first-order pattern can be seen below the image of torch. (b) Light diffracted from two adjacent grooves or marks, A and C, into the first-order pattern must have a path difference (AC–BC) of 1λ for constructive interference to occur

For light at grazing incidence and a first order pattern, n will be equal to 1. Using this knowledge it is possible to show that the formula for the wavelength of light is given by:

$$\lambda = sd^2 \,/\, 2D^2$$

where s is the spacing of the grating.

Using a steel rule with 0.5 mm divisions, at a distance D of 2 m from a pen torch, the middle of the first order will be found at a distance d of about 10 cm. Substituting:

$$\lambda = 0.5 \times 10^{-3} \times (0.10)^2 \,/\, 2 \times (2.0)^2$$
$$= 625 \text{ nm}$$

An accurate value would be 550 nm. Although the experimental result is not very close it is very instructive, especially as the whole experiment takes about 3 or 4 minutes to make, with very primitive equipment.

Much more accuracy could be obtained with a hi-fi record. Using a similar arrangement to that above, the value of d for the green part of the first order pattern will be observed to be near to 20 cm. The spacing of the grooves on the record can be measured with a low power microscope. In one experiment an average of nine grooves per mm was found. Substituting gives a λ value of 556 nm. It is surprising that such an accurate value can be obtained so easily.

6.4 DIFFRACTION BY CRYSTALS AND CRYSTAL-LIKE STRUCTURES

The atoms in crystals have a spacing similar to that of the wavelength of X-rays, and the diffraction of X-rays from these three-dimensional gratings has long been used for the elucidation of crystal structures. To be able to determine a crystal structure precisely it is necessary to measure the positions and intensities of the diffracted X-rays. However, even the *position alone* of a diffracted beam will give information about the spacing of the planes of atoms responsible. The well known formula relating these two aspects of diffraction is Bragg's law.

Consider the "reflection" of a beam (I) of monochromatic X-rays from a plane of atoms in a crystal, as sketched in Figure 6.6. Each atom acts as a scattering centre for the X-rays and the maximum intensity reflected will lie at the same angle θ with respect to the atom layer as the incident beam. (This is nothing more than the law of reflection.) If another beam (II) is reflected from a parallel layer of atoms a distance d below the first layer it will travel further than beam I. For beams I and II to reinforce each other they must be in phase on leaving the crystal. (Because the process occurring with both beams is identical we can ignore any change of phase that might occur on reflection.) This means that the path difference between beams I and II must be a whole number of wavelengths. Using the information in the figure it is seen that:

beam II has a longer path than beam I by CB + BD

For reinforcement

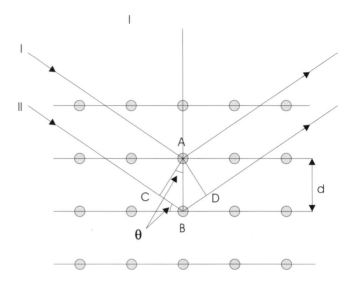

Figure 6.6 The geometry of the diffraction of X-rays from a crystal lattice needed to derive Bragg's law. Two beams, I and II, are "reflected" from two adjacent planes of atoms spaced a distance d apart and reinforce each other when the path difference between them is a whole number of wavelengths

$$CB + BD = n\lambda$$

where n is an integer and λ is the wavelength of the X-rays. However,

$$CB = BD = d\,\sin\theta$$

Hence

$$n\lambda = 2d\,\sin\theta$$

where d is the separation of the planes of atoms which are responsible for the diffraction, λ is the X-ray wavelength and θ the angle between the X-ray beam and the atom planes (which is *not* the conventional angle of incidence of optics) as shown in Figure 6.6. This relationship is Bragg's law. It was first applied in 1913 to determine the spacing of the lattice planes in a crystal of sodium chloride.

Although X-rays do not produce colours, the theory holds for any three dimensional crystalline array no matter the size of the "atoms". Thus any arrangement of particles, or even voids, which are spaced by distances similar to the wavelength of light will diffract light according to Bragg's law. When white light is used each wavelength will diffract at a slightly different angle and colours will be produced. Indeed many beautiful colours in nature are produced in this way.

As an example we can consider the gemstone precious opal. Common opal (potch opal) has a milky appearance and is the origin of the adjective opalescent. Precious opal shows flashes of brilliant colours from within the stone, seen over small angles as the stone is tilted. In the rarest opals the colours flash out from a black background.

The colour of precious opal is due to the diffraction of white light. The regions producing the colours are made up of an ordered packing of spheres of silica (SiO_2) which are embedded in amorphous silica or a matrix of disordered spheres, as illustrated in Figure 6.7. These small volumes resemble small crystallites. They interact with light because the spacing of the ordered regions of silica spheres is similar to that of the wavelength of light.

The conditions under which diffraction takes place are the same as those discussed with respect to the Bragg equation. However, it proves easier to model the effect by assuming that the diffracting centres are the small voids between the spheres, as drawn in Figure 6.8a. This indicates that the Bragg equation must be modified slightly because the diffraction does not take place in a vacuum, but within a silica matrix. It is thus necessary to use the optical path instead of the vacuum path. By analogy to previous discussions, d, the layer spacing of the voids (which is related to the diameter of the spheres) must be replaced by $n_s d$, where n_s is the refractive index of the silica in opal, about 1.45. The correct equation to use for opal is thus:

$$n\lambda = 2n_s d \sin\theta \sim 2.9\, d \sin\theta$$

Note also that refraction of the light beam will also take place at the opal surface and the diffraction angle, θ, will not be the same as the angle that the beam makes with the external surface.

When an opal is illuminated with white light some regions strongly diffract red, some green, and so on, dependent upon the angle of viewing and the void spacing or sphere diameter. Different orders of diffraction will be generated by different values of n, just as in the diffraction gratings discussed above. The colour of a single grain will change with viewing angle because of the $\sin\theta$ term in the Bragg equation. The longest wavelength observable, λ_{max}, will occur at normal incidence, when $\sin\theta$ is equal to one. In this case the wavelength diffracted back to the viewer will be:

ordered silica
sphere "crystallites"

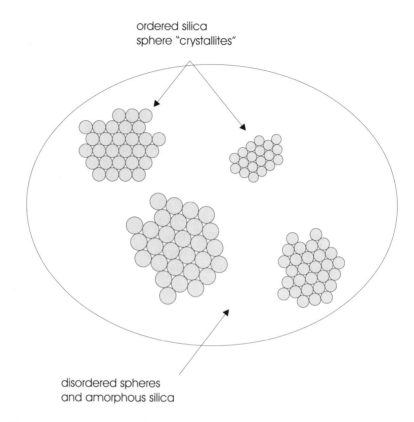

disordered spheres
and amorphous silica

Figure 6.7 The structure of precious opal consists of regions where spheres of silica pack together into ordered "crystallites" surrounded by a disordered matrix of silica spheres and amorphous silica. The ordered "crystallites" vary from one to another in orientation and in the diameters of the spheres involved

$$\lambda_{max} = 2n_s d \sim 2.9d$$

The relationship between the radius of the spheres, r, and the distance between the layers, d, will depend upon the exact geometry of the packing. If each layer of spheres is arranged in a hexagonal arrangement with each sphere touching each other, and successive layers are built up by placing further layers in the dimples of the layer below[3], the relationship between the sphere radius and the spacing of the voids will be:

---

[3] This arrangement is called closest packing. The two simplest ways of stacking the layers one on top of another are called hexagonal closest packing and cubic closest packing. Many more complex packing arrangements can be devised.

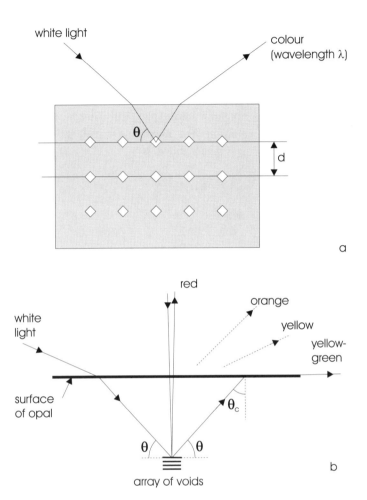

Figure 6.8 (a) Diffraction from precious opal can be modelled by diffraction from a series of voids (diamond shapes) in an amorphous silica matrix. When the spacing between the voids is given by d the wavelength diffracted, λ, will be given by 2.9 d sinθ. (b) Not all colours will be able to escape from the opal due to total internal reflection. If red light is observed normal to the diffracting layers, yellow-green light will just escape along the surface. All shorter wavelengths will remain within the opal

$$d = 2\sqrt{2}\,r / \sqrt{3}$$
$$= 1.633\,r$$

The radius of the spheres will be given by the maximum wavelength diffracted divided by 4.735. Because of the imprecision in the packing of spheres, a more useful general relationship is to say that the radius of the spheres is given, to a reasonable approximation, by one fifth of the wavelength of the colour observed at normal incidence.

As noted, as the diffracting grain is tilted the colour noted by a fixed observer will move from red towards violet. At a certain angle, however, total internal reflection will prevent the light from escaping, as shown in Figure 6.8b. The actual range of colour play will thus be less than that suggested by the Bragg equation. The maximum wavelength observed, at normal incidence, which was given above, is:

$$\lambda_{max} = 2.9d$$

At any other angle of incidence:

$$\lambda = 2.9\ d \sin \theta$$

If the silica has a refractive index of 1.45 and the opal has a flat surface and is surrounded by air, the critical angle (see Chapter 2), will be given by:

$$\sin\theta_c = 1.0 / 1.45$$
$$\theta_c = 43.6°$$

Reference to Figure 6.8b shows that if θ_c is equal to 43.6°, the diffraction angle, θ, is given by (90–43.6)°, that is 46.4°. Thus it is possible to write:

$$\lambda_{max} / \lambda_{crit} = 2.9d / 2.9 \sin \theta_c = 2.9d / 2.9d \sin 46.4°$$

where λ_{crit} is the wavelength of the colour diffracted just at the critical angle. Hence:

$$\lambda_{max} / \lambda_{crit} = 1 / \sin \theta_c = 1 / \sin 46.4° = 1.38$$

If the crystallite diffracts red light perpendicular to the surface, the light at a critical angle will have a wavelength of 507 nm, corresponding to yellow-green. No light of shorter wavelength will escape. Thus this plate-like opal will only show red to green colours when tilted.

6.5 Colour from Cholesteric Liquid Crystals

In Chapter 3 the structure of a liquid crystal made up of long molecules (a *nematic* liquid crystal) was described. In these mesophases, the *director*,

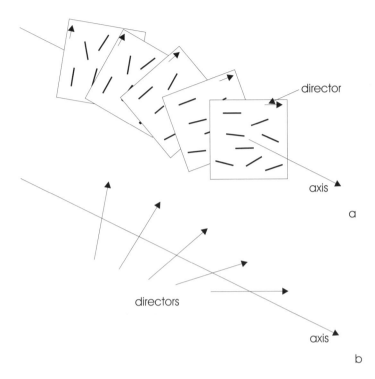

Figure 6.9 The cholesteric liquid crystal structure. (a) Layers of molecules and (b) the directors of each layer. The average orientation of the molecules (the director) in each layer rotates (by 20° in the diagram) in a regular fashion as one moves along the axis to create a helical structure

which is the average direction taken by the long axes of the molecules, adopts just one single direction. In *cholesteric* phases, which are also called *twisted nematic* phases or more correctly, *chiral nematic phases*, the director rotates steadily as one travels along a direction perpendicular to it. This is illustrated in Figure 6.9a,b. The result is the generation of a helical structure within the material. The physical reason for the rotation is that the molecules in each layer are asymmetric. When such molecules pack together the interactions are minimised if molecules in one layer rotate slightly compared to those in the layer below. Because these interactions are the same from one layer to another the same twist occurs between one layer to the next. In this way the helical structure results.

Colour from this array can arise by diffraction of light because the repeat structure of many cholesteric mesophases is similar to the wavelength of light. Thus light will be diffracted from the layers following Bragg's law:

$$n\lambda = 2nd \sin \theta$$

where n is the refractive index of the medium, d is the layer repeat distance and θ the angle between the layer and the light beam, exactly as in Figure 6.6. For light normally incident on the film:

$$\lambda = 2nd$$

When illuminated with white light, any wavelength satisfying the Bragg relationship will be diffracted strongly and give a colour in reflection. If the liquid crystal is backed by a dark background the colour will appear quite bright. On viewing the film normally and then moving towards grazing incidence the colour will appear to change towards shorter wavelengths due to the sin θ term in the Bragg equation, as discussed above.

The actual pitch of the helical structure and so the repeat distance of the layers can be engineered by both temperature and impurities. The end result is widely seen in liquid crystal thermometers. A commonly encountered form of these inexpensive devices consists of a card with a black strip of plastic running across it. In the black band a coloured number will be seen corresponding to the temperature. As the temperature varies a different number lights up at a different point on the band.

These devices operate in the following way. A series of spots of a cholesteric material are arranged in a row. They are chosen so that the periodicity of the molecular helix in each spot will diffract visible light at a precisely defined temperature. An increment of the temperature, $\delta T^\circ C$, of the order of 1°C is engineered between each successive spot by additives or by slightly modifying the cholesteric molecule. Within the design temperature range, each spot will diffract light only when the mesophase is at the correct temperature and not do so otherwise. This causes the appropriate temperature value to "light up". Moreover, on approaching the temperature, each spot will run through a spectrum of colours as the pitch of the helix varies. The effect of temperature on colour, of which this is an example, is called *thermochromism*.

These effects have already been anticipated by nature and a number of beetles show iridescent colours due to a cholesteric arrangement of layers of fibres in the outer integument of the body. The fibre direction in each layer is slightly different from that on either side and a helical layered structure is built up. If the pitch of the spiral is of similar dimensions to the wavelength of light this gives rise to intense "metallic" colours when viewed in white light. An example of metallic colours produced in this way are shown in Plate 6.1 of the bug *Zicrona callidea*.

6.6   ANSWERS TO INTRODUCTORY QUESTIONS

How can you use a steel ruler or a hi-fi record to measure the wavelength of light?
Either of these can be used as a reflection grating. If the light from a small pen torch is reflected from the grating at grazing incidence a coloured diffraction pattern looking like a spectrum will be seen as well as the image of the torch light. Measurement of the distance between the torch and the rule, and the apparent distance between the torch and the spectrum will allow the wavelength to be determined. A hi-fi record is better than a rule as it consists of a finer "grating".

Why are opals coloured?
Opals contain small volumes of ordered silica spheres packed together as in a crystal. The size of the spheres is appropriate for the diffraction of light. The colour diffracted depends upon the silica sphere size.

How do liquid crystal thermometers work?
Liquid crystal thermometers contain a line of spots of twisted nematic liquid crystals which possess a helical structure. The liquid crystals are chosen so that the repeat distance of the helix falls in the range of the wavelength of light. The temperature for which this is true is changed for each spot by slightly altering the properties of each liquid crystal. At an appropriate temperature a particular spot will diffract light and appear brightly coloured. At other temperatures the repeat distance of the twist will fall outside of the range and the spot will not diffract light. Thus the spots will light up in sequence along the line as the temperature varies.

6.7   FURTHER READING

The theory of diffraction is explained concisely by:

O. S. Heavens and R. W. Ditchburn, Insight into Optics, Wiley, New York (1991), Ch. 6.

The way to obtain the wavelength of light using a steel rule or similar grating and some informative background to the method is given in:

W. P. Trower (Ed.), Discovering Alvarez, University of Chicago Press, Chicago (1987), p. 1.

An introduction to the use of X-ray diffraction to determine crystal unit cells and as an aid to crystal identification is given by:

F. D. Bloss, Crystallography and Crystal Chemistry, Holt, Rinehart and Winston, New York (1971), Ch. 13.

The microstructure of precious opal is described in:

J. V. Sanders, Acta Crystallogr., **A24**, 1968, p. 427.
J. V. Sanders and P. J. Darragh, Mineralogical Record, **2**, No. 6, 1971.
P. J. Darragh, A. J. Gaskin and J. V. Sanders, Scientific American, **238**, April 1978, p. 87.

The cholesteric colours of certain beetles and how these may be proved to come from twisted layered structures is given by:

A. C. Neville and S. Caveney, Biol. Rev. **44**, 1969, p. 531.

6.8 PROBLEMS AND EXERCISES

1. Calculate the width of a slit needed to give a first maximum at about an angle of 0.5° for light of wavelength 550 nm.

2. An astronomical telescope has a mirror of diameter 100 cm and a focal length of 2000 cm. Calculate the diameter of the first bright ring surrounding the image of a star in the focal plane for light of wavelength 550 nm.

3. Derive the formula

$$\lambda = sd^2 / 2D^2$$

where s is the spacing of a reflection diffraction grating and the other distances are given in Figure 6.5. Use the fact that the path difference for constructive diffraction between rays from adjacent lines on the grating must be equal to 1λ.

4. A hi-fi record is used as a grating in the arrangement shown in Figure 6.5. The position of a red band was found at 52 cm below the undeviated position when viewed from a distance of 750 cm. If the wavelength of the light is taken as 650 nm what is the spacing of the record grooves?

5. In an X-ray diffraction experiment, a strong reflection corresponding to planes of atoms separated by half of the cubic unit cell edge of nickel oxide, NiO is observed at a θ value of 21.66° when copper radiation of wavelength 0.15418 nm is used. Calculate the length of the cubic unit cell.

6. The structure of magnesium consists of layers of Mg atoms packed on top of each other. Each layer is made up of a close packed hexagonal array of atoms. If the X-ray reflection from two adjacent layers, taken with copper radiation of wavelength 0.15418 nm is at an angle θ of 17.25°, calculate the diameter of a magnesium atom.

7. Estimate the diameter of the silica spheres giving rise to green reflections in opal when the gemstone is viewed from directly above.

8. When viewed from above, a liquid crystal thermometer is orange. What is the layer repeat spacing of the liquid crystals? The refractive index of the liquid crystals is 1.488. When viewed at an angle of approximately 57° to the vertical the colour becomes green. Explain this.

CHAPTER 7

COLOUR FROM ATOMS AND IONS

How can "Neon" signs be made to show a variety
of colours?
How do sodium street lamps produce yellow light?
Why are transition metal compounds coloured?
Why is ruby red and copper sulphate blue?

7.1 THE SPECTRA OF ATOMS AND IONS

We saw in Chapter 1 that the spectrum emitted by an incandescent solid
was a continuous spectrum. The spectrum of a rarefied gas of atoms or ions
consists of discrete sharp lines. If the light is emitted by the gas a sequence
of bright lines is observed. If white light is passed through a gas the
spectrum will show dark lines on the continuous bright background. In this
chapter the origin of these lines and the colours that they give rise to is
described.

For most chemical purposes an atom or an ion can be considered to
consist of a dense minute nucleus surrounded by electrons. In this and
following chapters we will be concerned with the electrons, especially those
of highest energy. They are thought of as having two types of structure
within an atom, a pattern of energy level occupation and a pattern of space
occupation. The terminology is given in Appendix 7.1 and 7.5.

Electron energy levels in isolated atoms and ions are sharp. Energy is
absorbed when electrons are excited from a lower energy level to a higher
level and exactly the same energy is released when the electron drops back
to the same lower level again. Although most electron transitions involve
high energy changes such that the radiation absorbed or emitted is in the
X-ray and ultraviolet regions, in some atoms there are electron transitions
where the energy change is smaller. The radiation absorbed or emitted is in
the visible and hence these transitions can give rise to colours.

When a photon of energy $h\nu$ is *absorbed* by an atom or ion it passes from
a lower energy state, often the lowest available state in the system, called
the *ground state*, to an upper one, as shown in Figure 7.1a. The transition
will take place if the frequency of the photon, ν, is given exactly by:

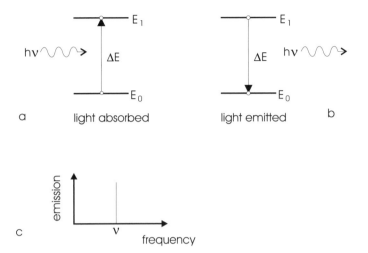

Figure 7.1 The absorption and emission of radiation by isolated atoms or ions. (a) Light is absorbed if the energy of the photon exactly matches the energy gap, ΔE, between the lower level and the upper level. (b) When energy is released spontaneously the energy of the photon is again exactly equal to ΔE, the difference in energy between the upper and lower levels. (c) The emission from a collection of atoms in a gas at low pressure will consist of a sharp line. The frequency at which the line occurs is the same as that of the photons involved in absorption and emission in (a) and (b)

$$\nu = (E_1 - E_0) / h = \Delta E / h$$

where E_0 is the energy of the lower or ground state, E_1 is the energy of the upper state and h is Planck's constant. If the atom is in the upper state, E_1, and makes a transition to the lower state, E_0, energy will be emitted, as illustrated in Figure 7.1b. This will have the same frequency, given by the same equation:

$$\nu = (E_1 - E_0) / h = \Delta E / h$$

Each transition gives rise to a sharp line in the spectrum, depicted in Figure 7.1c. For example, if the light from a mercury lamp or a sodium lamp is dispersed by a diffraction grating or a prism[4], instead of the familiar continuous spectrum discussed in Chapter 2, one will see a series of sharp

---

[4] Devices for the display of spectra are called spectroscopes, spectrographs or spectrometers. They use prisms or diffraction gratings and the resultant spectra are recorded and displayed electronically.

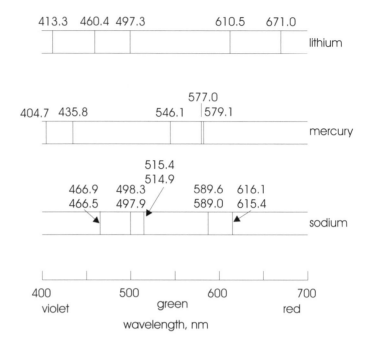

Figure 7.2 Schematic depiction of the line spectra from atoms of lithium, mercury and sodium. The wavelength of each line is noted in nm. The sodium spectrum consists of very closely spaced pairs of lines (doublets), represented here as single lines

lines. This pattern is called a *line spectrum*. Some examples are illustrated in Figure 7.2.

7.2 ATOMIC SPECTRA AND CHEMICAL ANALYSIS

The interaction between electrons and radiation is governed by quantum mechanical rules which state: (i) light can only interact if it has exactly the correct energy to allow the electron to pass from one well defined energy level to another; (ii) light can only interact if the wave functions specifying the two states are suitable. This latter condition leads to a number of *selection rules* which allow one to determine whether the transition is probable or improbable.

The exact arrangement of the energy levels in an atom is very sensitive to changes in the electron configuration. When this is coupled to the selection rules operating, it emerges that the line spectrum of each

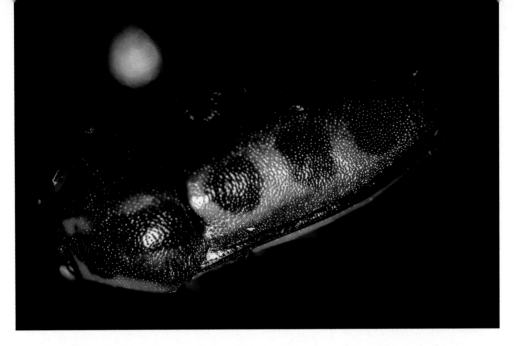

Plate 6.1 Iridescent colours displayed by a bug (*Zicrona callidea*). These arise from a helical arrangement of the layers that make up the outer integument of the insect with a pitch of the order of the wavelength of light. I am indebted to Dr M. Wilson of the Museums and Art Galleries of Wales, Cardiff, for the loan of specimens of *Zircrona callidea*

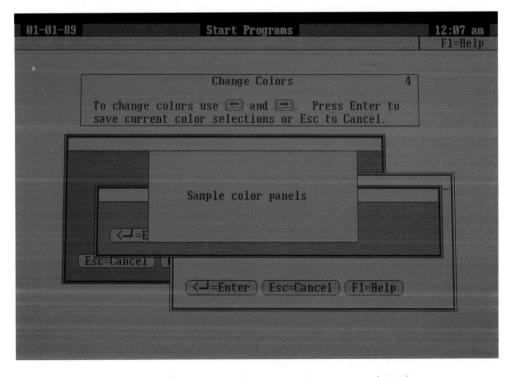

Plate 7.1 A gas plasma flat screen display in a portable computer of 1989

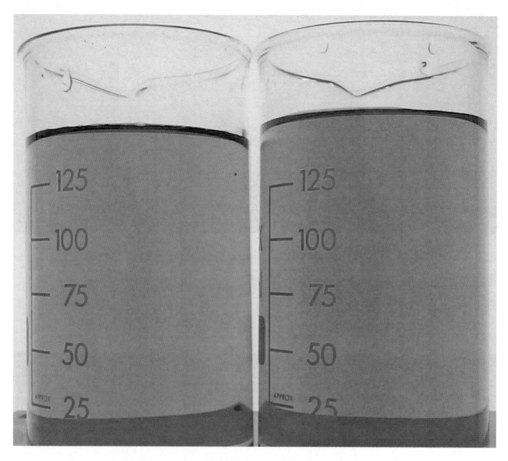

Plate 7.2 Crystal field colours of transition metal ions. (a) The green colour of $Ni(H_2O)_6^{2+}$ ions and the blue colour of $Cu(H_2O)_6^{2+}$ ions in water solution

(b) Crystals of the compounds $Ni(NO_3)_2.6H_2O$ (green) and $Cu(NO_3)_2.3H_2O$ (blue), showing that the transition metal ions are in similar environments in both the crystals and solutions

Plate 7.3 The differing crystal field colours of Cu^{2+} ions in malachite, $(Cu_2(OH)_2(CO_3)$, green) and azurite $(Cu_3(OH)_2(CO_3)_2$, blue) intermingled in a mineral sample

Plate 7.4 A glass bottle coloured blue by the addition of Co^{2+} ions, which occupy tetrahedral sites in the glass matrix

Plate 7.5 An enamel trinket box lid. The enamel is a glass-based material which has been fused to a metallic base. The colours derive from oxide pigments dissolved or dispersed in the glass

Table 7.1 Flame test colours

Atom	Colour	Atom	Colour
Lithium[a]	Scarlet		
Sodium	Yellow	Calcium	Orange-red
Potassium	Violet	Strontium	Crimson
Rubidium	Red-violet	Barium	Yellow-green
Caesium	Blue		

[a] The colour of lithium arises from LiOH molecules rather than isolated Li atoms.

chemical element is unique. Thus the spectrum becomes a powerful analytical tool. Each atom or ion can be thought of as having a line spectrum fingerprint which can be used as a diagnostic test for the element.

At the simplest level this is made use of in inorganic chemistry as a "flame test". A small quantity of the material being examined is placed upon a platinum wire and heated to high temperature in a flame. The colour of the flame is often a good guide to the metal present. This method works well with the alkali metals and alkaline earth metals, which produce the colours given in Table 7.1.

Much more information about the colours given out by the flame can be obtained by allowing the light to pass through a narrow slit and viewing it with an inexpensive plastic diffraction grating. The grating spreads the light out into a series of spectra which in this case consists mainly of lines. In fact such an arrangement is a simple spectroscope. The technique can yield more information if the intensities and positions of the lines in the spectrum can be recorded. Comparison of these intensities with those from standard solutions of ions allows quantitative analysis of even very small quantities of impurities to be made. The technique is called *atomic absorption* analysis. It is routinely used to detect quantities of metal impurities at concentrations of parts per million.

7.3 FRAUNHOFER LINES AND STELLAR SPECTRA

In Chapter 1 we noted that the spectrum of an incandescent body such as the sun would be a continuous spectrum. In 1814, Joseph Fraunhofer, by making better spectrographs than any others available at that time, discovered that the solar spectrum was interspersed with a number of dark lines, now called *Fraunhofer lines*. The most important visible Fraunhofer lines are illustrated in Figure 7.3 and listed in Table 7.2. These features are actually *absorption spectra* and consist of both sharp lines and wider

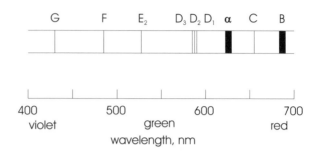

Figure 7.3 The main visible Fraunhofer lines and bands. The atoms or molecules responsible for these features are listed in Table 7.2

Table 7.2 Some Fraunhofer lines

Designation	Origin	Wavelength (nm)
B	O_2 molecules	687.7–688.4 (band)
C	Hydrogen	656.2
α	O_2 molecules	627.6–628.7 (band)
D_1	Sodium	589.6
D_2	Sodium	589.0
D_3	Helium	587.6
E_2	Iron	527.0
F	Hydrogen	486.1
G	Iron, calcium	430.8

bands. The lines are due to absorption by single isolated atoms and ions while the bands arise from molecules. The origin of these bands is described in the following chapter.

The Fraunhofer lines arise in two ways. One set of absorption lines, the *telluric* lines, are due to components of the Earth's atmosphere. These absorb incoming solar radiation and give rise to dark lines or bands in the otherwise continuous spectrum from the sun. The principal contributions are from oxygen and water vapour. However, another set of lines arises when light from the sun is absorbed by atoms or ions in the relatively cool outer solar regions. Among the most prominent of these are lines from hydrogen, sodium, calcium and iron, which could all be identified by comparison with spectra from standards available in the laboratory. Significant among the information, a Fraunhofer line at a wavelength of 587.6 nm, discovered in 1868, could not be attributed to any known element. The new line was taken as an indication of the presence of a new element in the solar atmosphere. The element was subsequently named *helium*, from the

Greek word for the sun, helios. Almost 30 years were to pass before the gas was discovered on the Earth, by Ramsay, who isolated it in 1895. Some of the most intense visible Fraunhofer lines are listed in Table 7.2.

Nowadays the presence of metallic atoms and ions in the outer atmospheres of stars or even far off galaxies is generally confirmed by recording the spectrum of the star and examining the dark absorption Fraunhofer lines found.

7.4 NEON SIGNS AND STREET LIGHTS

Faraday, in 1835, first discovered that gases at low pressure could conduct electricity and at the same time give out light. The complex processes taking place were investigated in depth by Geissler in the 1860s and at the end of the 1890s by Crookes.

The experimental observations are easy to report. If a gas is contained in a tube and is at atmospheric pressure it will not conduct electricity. If the pressure is reduced and the gas is subjected to a high voltage (of the order of kilovolts) it begins to show electrical conductivity and at the same time it starts to emit light. The colour of the light depends upon the gas in the tube. This is the basis for the operation of neon signs, invented in 1910, and of sodium and mercury vapour street lighting.

In these devices, electrons are emitted from the cathode (the negative electrode) and are accelerated in a high electric field across the gas. These electrons collide with gas molecules and excite them to higher energies. Some molecules are ionised and these ions in turn are accelerated in the electric field and cause further ionisation and excitation. The light emitted is due to the excited atoms and ions losing energy by releasing photons as they return to lower energy states.

"Neon" signs make use of the group of elements known as the inert gases. These elements exist as monatomic gases at normal temperatures and all can be used in what are now collectively known as neon signs, neon being the first to be used. To make a neon sign a glass tube is evacuated and filled with a low pressure of one of the inert gases. The gas is subjected to a voltage of about 10 kV via electrodes at opposite ends of the tube. The outer electron configuration of the inert gas group of atoms and the colours that they produce in neon signs are given in Table 7.3.

With the advent of portable computers a need arose for a lightweight flat display screen. Among the first of these was the monochrome gas plasma display, which operated on the principles just outlined. Ionised inert gases

Table 7.3 The colours produced by the inert gases

Gas	Outermost electron configuration	Colour
Helium	$1s^2$	Yellow
Neon	$2s^2, 2p^6$	Pink-red
Neon + argon	–	Red
Argon	$3s^2, 3p^6$	Pale blue
Argon + mercury	–	Blue
Krypton	$4s^2, 4p^6$	Lavender
Xenon	$5s^2, 5p^6$	Blue

(mainly neon) were employed to produce the illumination. A photograph of a display of this type, from a computer available in 1989, is shown in Plate 7.1. This technology rapidly gave way to full colour displays which are described in Chapter 12.

Sodium street lights give out a characteristic yellow colour, which arises from excited sodium atoms. As sodium is a solid at normal temperatures the discharge is started via a low pressure of neon which is also contained in the lamp tube. This "neon lamp" is first activated, which is the reason why sodium lamps glow with a pink-red colour when they are warming up. After a short time the energy supplied to the neon generates enough heat for the sodium to evaporate. At this stage electronic and atomic collisions involving sodium produce the well known yellow sodium light.

A partial energy level diagram of sodium atoms is given in Figure 7.4. The energising collisions excite the sodium to the higher energy levels, from which they return to the ground state by releasing photons. Each return path passes through the energy levels marked $3^2P_{1/2}$ and $3^2P_{3/2}$ before returning to the ground state $3S_{1/2}$. It is these two last transitions which give out the familiar yellow light. The two 3^2P levels differ in energy by 0.0032×10^{-19} J so that the yellow sodium light consists of two wavelengths, 588.995 nm and 589.592 nm. These constitute the *sodium D lines*, widely used in spectroscopy and as a standard wavelength at which to record optical properties such as refractive index. The line spectrum of sodium is illustrated in Figure 7.2.

Mercury vapour lights operate in a similar fashion to sodium lights. A schematic diagram of the important energy levels is reproduced in Figure 7.5. It can be seen from the figure that there are a number of transitions with wavelengths in the blue-green, which gives these lamps their rather eerie coloration. The visible line spectrum of mercury is shown in Figure 7.2. However, the strongest emission is from the 6^3P level to the ground

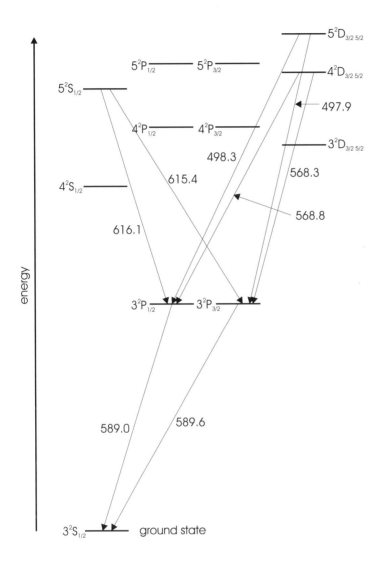

Figure 7.4 Schematic energy level diagram for sodium atoms. The transitions giving rise to colours are drawn as downward pointing arrows. The wavelengths emitted are given besides each transition (in nm). The transitions at 589.0 and 589.6 nm produce a bright yellow doublet; the 615.4 and 616.1 nm transitions are dim red; the 568.8 and 568.3 nm transitions are dim yellow and the 498.3 and 497.9 nm transitions are dim green. The states labelled 3^2D, 4^2D and 5^2D each consist of a pair of closely spaced energy levels

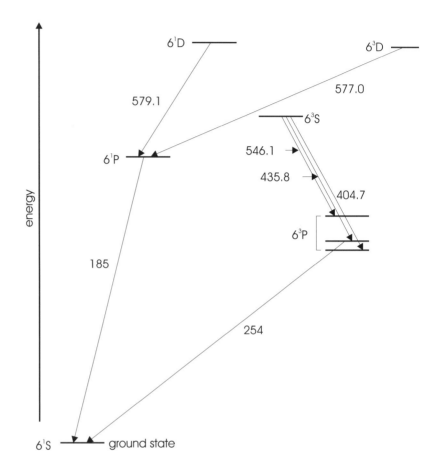

Figure 7.5 Schematic energy level diagram for mercury atoms. The transitions giving rise to colours and the two important ultraviolet transitions are marked by downward pointing arrows. The wavelengths emitted are given besides each transition (in nm). The bulk of the emission (65%) is in the ultraviolet at 254 nm. The level labelled 6^3D consists of three closely spaced levels

state 6^1S, producing ultraviolet radiation with a wavelength of 254 nm. To make use of this output mercury tubes are often coated with a fluorescent material which converts this ultraviolet component into visible light, as explained in Chapter 9.

The line spectra of street lights are, in fact, easy to observe or photograph using an inexpensive plastic transmission diffraction grating. View the light through the grating at the correct angle to see the first order pattern. A series of lines will be observed, that are quite different from the continuous spectrum given out by, for example, a torch. More information will be found in Section 7.13.

7.5 TRANSITION METALS AND CRYSTAL FIELD COLOURS

Generally electron transitions in atoms do not produce visible radiation. Exceptions are the *transition metal* ions, which are often described as "coloured". The most important transition metals from the point of view of colour are the 3d transition metals, listed in Appendix 7.2. As an example of these colours, Plate 7.2a,b shows aqueous solutions and crystals of green $Ni(H_2O)_6^{2+}$ and blue $Cu(H_2O)_6^{2+}$. In both of these ions, the six water molecules are arranged so that the oxygen atoms form an octahedral co-ordination polyhedron around the central cation. The colours can be quantified by recording the absorption spectra of the solutions. These are reproduced in Figure 7.6a,b and show that the nickel containing solution absorbs in both the violet and red regions of the spectrum while the copper containing solution absorbs only in the yellow to red region.

In these and the other 3d transition metal ions the 3d orbitals contain one or more electrons and it is these that are associated with the colours observed. Electron transitions between the various d-orbitals produce colours. This comes about in an interesting way and involves the *shapes* of the d-orbitals. The d-orbitals point along or between the x-, y- and z-axes as can be seen in Figure 7.7. The orbitals directed between the axes, d_{xy}, d_{yz}, d_{xz}, are called the t_{2g} or t_2 set and those pointing along the axes, $d_{x^2-y^2}$ and d_{z^2}, are called the e_g or e set. In an isolated transition metal atom or ion the d-orbitals all have the same energy. When the atom or ion is placed into a crystal the energy of the d-orbitals increases, due to repulsion between the surrounding electrons and the electrons in the d-orbitals. If these surrounding electrons were distributed evenly over the surface of a sphere the five d-orbitals would still have the same energy as each other, although much higher than in the isolated state. However, it is not difficult to imagine that if the surrounding electrons are arranged differently the energy of some of the d-orbitals might be different than the others. This is called *crystal field splitting* or *ligand field splitting*.

The effect can be illustrated by considering a transition metal ion surrounded by an octahedron of negative O^{2-} ions, as in Figure 7.8a. The d-orbitals pointing directly towards the oxygen ions, $d_{x^2-y^2}$ and d_{z^2}, (the e_g pair) will be strongly repelled and so raised in energy compared to those pointing between the oxygen ions, the d_{xy}, d_{xz} and d_{yz} (the t_{2g}) group. The

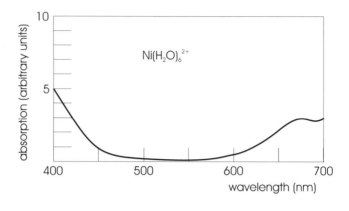

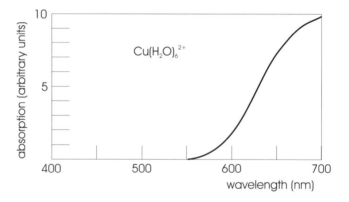

Figure 7.6 The absorption spectra of aqueous solutions containing (a) $Ni(H_2O)_6^{2+}$ and (b) $Cu(H_2O)_6^{2+}$. The intensity scales are arbitrary

result is shown on the right hand side of Figure 7.9. The crystal field splitting generates an energy gap between the lower t_{2g} group of orbitals and the upper e_g group which is written as Δ.

When a transition metal ion is surrounded by a tetrahedron of O^{2-} ions, as in Figure 7.8b, the crystal field splitting is reversed. In this case the d_{xy}, d_{xz} and d_{yz} orbitals (the t_2 group) are raised in energy relative to $d_{x^2-y^2}$ and d_{z^2} (the e pair). The magnitude of the splitting for ions in a tetrahedron will be less than that for ions in an octahedron because there are only four negative ions instead of six and because they are further away from the central ion. Calculations give the result that the tetrahedral crystal field splitting is 4/9 of the octahedral splitting, as indicated on the left-hand side of Figure 7.9.

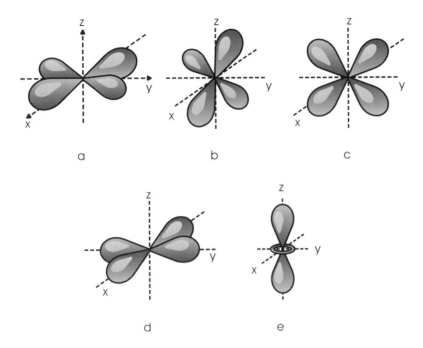

Figure 7.7 The shapes of the five d-orbitals superimposed upon a set of orthogonal axes: (a) d_{xy}; (b) d_{xz}; (c) d_{yz}; (d) $d_{x^2-y^2}$; (e) d_{z^2}. The lobes of the group (a, b, c) all lie between the axes and are collectively called the t_2 or t_{2g} set. The lobes of the pair of orbitals (d, e) lie along the axes and are collectively called the e or e_g set

The colour produced by an ion is due to d-electrons moving across the energy gap created by the crystal field splitting. For example, in the case of $Cu(H_2O)_6^{2+}$, shown in Plate 7.2, the value of the octahedral crystal field splitting, Δ, produced by the surrounding oxygen atoms on the water molecules, is 2.58×10^{-19} J (1.61 eV). A transition of this energy produces an absorption peak with a maximum wavelength near to 770 nm in the near infrared. It is the high energy tail of this peak, visible in Figure 7.6b, that gives the solution its blue colour.

The magnitude of the crystal field splitting will depend on the geometry of the surrounding ions and how close they are to the cation. In a strong crystal field, produced when the surrounding anions are close to the cation, the crystal field splitting is large. In a weak crystal field, produced when the surrounding anions are further away from the cation, the splitting is smaller. This variation accounts for the fact that any particular transition metal cation may exhibit different colours in different compounds.

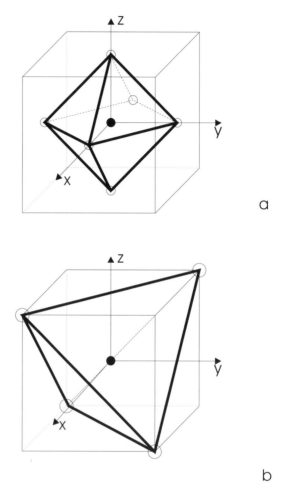

Figure 7.8 (a) A cation surrounded by six oxygen ions arranged as an octahedral anion coordination polyhedron. (b) A cation surrounded by four oxygen ions arranged as a tetrahedral anion coordination polyhedron. The cations are shown as filled circles and the anions as open circles. The reference cubic outline indicates that the cation–anion distance is greater in (b) than (a)

In order to relate these ideas to the energy level diagrams shown in Figures 7.4 and 7.5 we need to construct the energy levels of the transition metal ions in the crystal field. The calculation of these is complex and we will not go into detail here. However, the general idea of the process can be easily explained. Suppose that we have a d^3 ion in an octahedral site in an oxide, such as Cr^{3+} in ruby. The lowest energy for the ion will be when all

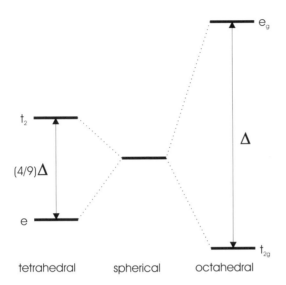

Figure 7.9 The response of the five d-orbitals to a surrounding crystal field. In a spherical field the energies of the orbitals are the same. In an octahedral field the t_{2g} set have a lower energy and the e_g set a higher energy with respect to the spherical group, while in a tetrahedral field the situation is reversed. The separation of the upper and lower energy levels, the crystal field splitting, is Δ in the case of an octahedral crystal field and $(4/9)\Delta$ for a tetrahedral crystal field. Colour is produced by electron transitions between the lower and upper groups of orbitals

three d electrons are in the lowest energy state, $(t_{2g})^3 (e_g)^0$, as in Figure 7.10a. The next lowest energy for the ion will correspond to two electrons in the t_{2g} orbitals and one in the e_g pair, $(t_{2g})^2 (e_g)^1$ depicted in Figure 7.10b. Another energy level will correspond to the electron distribution, $(t_{2g})^1 (e_g)^2$, Figure 7.10c. Finally, the highest energy will correspond to all electrons in the upper level, $(t_{2g})^0, (e_g)^3$, as in Figure 7.10d. We will end up with four energy levels for the Cr^{3+} ion, set out schematically in Figure 7.10e. This is the situation which is found in a *strong crystal field*, when all other interactions can be ignored.

If the crystal field is *weak* other interactions cannot be ignored. Of these, electron–electron repulsion is most important. If the crystal field splitting is ignored and electron–electron repulsion is considered to be dominant a set of energy levels described by "term schemes" can be derived (see Appendix 7.4). In most real crystals the crystal field splitting, while not being strong in the sense defined above, cannot be ignored. The energy levels are then somewhere between the weak field and strong field extremes and are often labelled with respect to the group theoretical

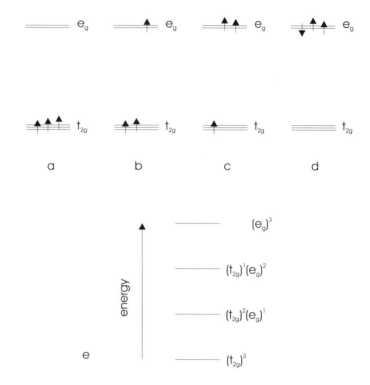

Figure 7.10 The possible electron distributions between the t_{2g} and e_g orbitals in an octahedrally coordinated Cr^{3+} ion (a–d) leads to four energy levels (e). Note that each of the three orbitals in the t_{2g} and each of the two in the e_g group have the same energy. They have been separated for clarity

symmetry of the species. The precise explanation of the energy levels and colour of green $Ni(H_2O)_6^{2+}$ mentioned above and shown in Plate 7.2, and ruby, described below, require this more elaborate approach.

7.6 THE COLOUR OF RUBY

Ruby consists of single crystals of *corundum*, Al_2O_3, containing about 0.5% Cr as an impurity. The Cr^{3+} ions are distributed at random over some of the positions normally reserved for Al^{3+} in the oxide lattice. These positions are *octahedral sites* as the oxygen ions lie at the vertices of a slightly distorted octahedron around the cations. The fact that ruby is coloured while Al_2O_3 is colourless indicates that it is the Cr^{3+} in the structure that is of paramount

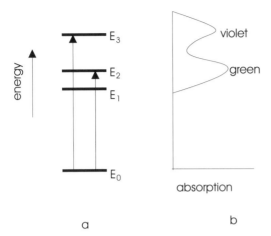

Figure 7.11 The energy levels of Cr^{3+} in ruby. (a) The Cr^{3+} ions introduce three energy levels, E_1, E_2 and E_3, above the ground state, E_0. Transitions from the ground state to levels E_2 and E_3 produce the absorption spectrum shown in (b). The colour of the gemstone is due to the non-absorbed colours, i.e. red with a slightly blue tint

importance. It is thus necessary to focus attention upon the energy levels of the Cr^{3+} ions themselves. The most significant energy levels, which derive from the combined effect of the crystal field splitting and electron–electron interactions, are shown schematically in Figure 7.11a.

When white light falls on the ruby crystal, energy is absorbed in exciting the ion from the ground state, E_0, to the two upper levels E_2 and E_3. Because of interactions with the surroundings the absorption spectrum consists of two rather broad peaks, as sketched in Figure 7.11b. The transition from E_0 to level E_2 absorbs light with a wavelength centred upon 556 nm, in the yellow/green region of the spectrum. The transition from E_0 to E_3, absorbs light with a wavelength centred upon 407 nm, close to violet. The light transmitted or reflected is depleted in these colours and appears red with a slight blue cast, the typical colour of ruby.

The crystal structure of ruby is hexagonal and as with all crystals of symmetry lower than cubic the absorption spectrum depends upon the polarisation of the light used for the illumination. In ruby two absorption spectra arise, one for light polarised parallel to the crystallographic c-axis and one for light polarised perpendicular to the c-axis (the optical axis). Although these spectra are very similar to each other, as can be seen from Figure 7.12, noticeable differences in colour are apparent when ruby crystals in differing orientations are observed in polarised light. We have

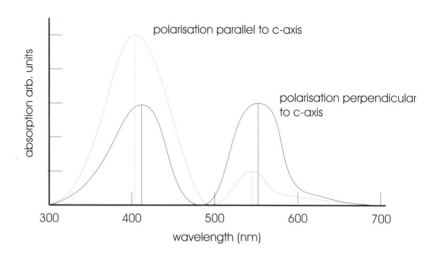

Figure 7.12 Absorption spectra of ruby taken using light polarised parallel and perpendicular to the crystallographic c-axis (the optical axis) of the crystal. The differences between the spectra cause the colour of the crystal to change with orientation when viewed in polarised light, the phenomenon of dichroism

met with this phenomenon, *dichroism*, in Chapter 3. In ruby, when the plane of polarisation of the light is perpendicular to the c-axis the crystal is perceived as ruby coloured but when it lies parallel to the c-axis the crystal takes on a more orange hue.

7.7 THE COLOUR OF EMERALD

The effect of the strength of the crystal field is demonstrated by comparing ruby with *emerald*. Emeralds possess the *beryl* ($Be_3Al_2Si_6O_{18}$) structure, but, like ruby, also contain some chromium. In emerald Cr^{3+} is also found in octahedral sites, replacing some of the Al^{3+} ions present However, in beryl the octahedra surrounding the Cr^{3+} ions are slightly larger than in corundum and so the crystal field experienced by the Cr^{3+} in emerald is weaker than in ruby. The energy level diagram remains essentially the same, but there is a shift in the energy levels E_2 and E_3 towards the ground state level E_0. This causes the two main absorption bands (Figure 7.11) to move towards the lower energy red end of the spectrum. The band that absorbs yellow/green in ruby now absorbs yellow/red. The violet absorbing band in ruby now absorbs more blue. Emeralds therefore absorb red and some blue and transmit green with some residue of blue to give the typical emerald colour.

7.8 CRYSTAL FIELD COLOURS IN MINERALS AND GEMSTONES

Although the colour produced by a transition metal ion will depend on the local crystal field many transition metal ions are thought of as showing a typical colour. Copper compounds, for example, are usually green-blue and Fe^{2+} imparts a pale watery green colour to oxides and hydrates. However, this generalisation must be treated with considerable caution. Many ions can occur in more than one type of coordination and important variations of colour can occur. The colour of Cr^{3+} in octahedral sites in ruby and emerald has already been cited. When this ion is incorporated into chrome alum the colour perceived is purple. Plate 7.3 shows a sample of the mineral *malachite*, $Cu_2(OH)_2(CO_3)$, with a green colour. The bands indicate that the mineral was laid down as a sedimentary rock over aeons of time. This is mixed with the related mineral *azurite*, $Cu_3(OH)_2(CO_3)_2$, which is bright blue. Although the chemical formulae of these two minerals is similar, the crystal field in azurite is sufficiently different to that in malachite that quite different colours are perceived, blue versus green. Despite this limitation, it is useful to list some of the more characteristic colours of the 3d transition metal compounds. These are given in Table 7.4.

As was highlighted by ruby and emerald, many gemstones are actually coloured by the presence of small amounts of transition metal impurities in what would otherwise be colourless crystals. A short list of some gemstones coloured by transition metal ion impurities is given in Table 7.5.

7.9 COLOUR AS A STRUCTURAL PROBE

The spectrum of a transition metal ion in a solid can give information about the local position of the ion because the colour depends upon the crystal field. Consider as an example the problem of cation distribution in oxide spinels.

The spinel structure is adopted by many compounds with a formula AB_2X_4, where A and B are medium sized cations and X represents an anion, most often O^{2-}. In this structure the anions are in a close packed array and the cations sit in octahedral and tetrahedral sites. The absorption spectra and colour of transition metal ions are quite different for these geometries and so the site occupation can be easily and unambiguously determined. The spinel $NiAl_2O_4$ is a case in point. The absorption spectrum of this

Table 7.4 Typical colours shown by 3d transition metal compounds

Ion	d electrons	Colour	Example
Ti^{3+}	1	Purple	Glass[a]
V^{4+}	1	Red	Glass
V^{3+}	2	Green	Glass
		Blue	Al_2O_3
Cr^{3+}	3	Green	Glass, emerald ($Be_3Al_2Si_6O_{18}$)
		Red	Al_2O_3 (ruby), TiO_2
		Violet-purple	Chrome alum ($KCr(SO_4)_3.12H_2O$)
Mn^{3+}	4	Purple	Glass
Mn^{2+}	5	Yellow	Glass
		Red	$MnCO_3$
		Green	MnO
		Pink	$MnSiO_3$
Fe^{3+}	5	Yellow-green	Glass
Fe^{2+}	6	Blue-green	$Fe(H_2O)_6^{2+}$ in solution and hydrates
Co^{2+}	7	Blue	$CoAl_2O_4$, glass
		Pink	$Co(H_2O)_6^{2+}$ in solution and hydrates
Ni^{2+}	8	Green	$Ni(H_2O)_6^{2+}$ in solution and hydrates
		Yellow	Al_2O_3
Cu^{2+}	9	Green	$Cu_2(OH)_2(CO_3)$ (malachite)
		Blue	$Cu(H_2O)_6^{2+}$ in solution and hydrates

[a] Glass refers to ordinary soda-lime silicate glass.

material reveals that the Ni^{2+} ions are found in both positions. Moreover, spectra taken in the earliest stages of the formation reaction:

$$NiO + Al_2O_3 \rightarrow NiAl_2O_4$$

shows that the Ni^{2+} occupies both sites from the very start of the reaction. In the related spinel $NiGa_2O_4$, the Ni^{2+} ions exclusively occupy octahedral sites.

Another difficult problem relates to the cation coordination in glasses. X-ray crystallography, the traditional way of solving these structural questions, fails with amorphous materials. However, it is possible to incorporate a small amount of a transition metal into the structure as a probe of local geometry. For example, Plate 7.4 shows a glass bottle incorporating a small quantity of Co^{2+}. The blue colour is considered to be typical of tetrahedrally coordinated Co^{2+} in oxide matrices and indicates that the medium sized Co^{2+} ions replace the small Si^{4+} ions in tetrahedral sites in the glass network.

The tetrahedral network structure of silicate glass is well known, but new glasses often have quite unknown structures. As an illustration of the use of

Table 7.5 The colours of some gemstones

Gemstone	General formula	Structure type	Colour	Origin of colour and cation replaced
Garnet	$Ca_3Al_2Si_3O_{12}$	Garnet	Red	Fe^{2+} in cubic (8-coordinate) Ca^{2+} site
Peridot	Mg_2SiO_4	Olivine	Yellow-green	Fe^{2+} in octahedral Mg^{2+} site
Topaz	$Al_2SiO_4(OH,F)_2$	–	Yellow	Fe^{3+} in octahedral Al^{3+} site
Emerald	$Be_3Al_2Si_6O_{18}$	Beryl	Green	Cr^{3+} in octahedral Al^{3+} site
Alexandrite	$BeAl_2O_4$	–	Red/Green	Cr^{3+} in octahedral Al^{3+} site
Ruby	Al_2O_3	Corundum	Red	Cr^{3+} in octahedral Al^{3+} site
Tourmaline	$CaLi_2Al_7(OH)_4(BO_3)_3Si_6O_{18}$	–	Pink-red	Mn^{2+} in octahedral Al^{3+} site
Turquoise	$CuAl_6(PO_4)_4(OH)_8 \cdot 4H_2O$	–	Blue-green	Cu^{2+} in octahedral (Cu^{2+}) site

transition metal ions as structural probes we can consider an exploration of the structure of a $ZnCl_2$ glass. A small amount of Mn^{2+} incorporated into a $(Mn,Zn)Cl_2$ glass imparts a yellow colour and yields a spectrum typical of tetrahedrally coordinated Mn^{2+}. As with the silicate glasses, small additions of Co^{2+} in $(Co,Zn)Cl_2$ gives a blue material characteristic of tetrahedrally coordinated Co^{2+} ions. These results give strength to the argument that amorphous $ZnCl_2$ is formed from a random network of linked $ZnCl_4$ tetrahedral units and that the added ions replace Zn^{2+} in the network. The suggestion is further strengthened by noting that incorporation of Fe^{2+} in $(Fe,Zn)Cl_2$ glass yields an absorption spectrum expected from tetrahedrally coordinated Fe^{2+} ions.

This measurement of absorption spectra can also be used to determine the oxidation states of transition metal ions and indirectly to yield information upon the local conditions prevailing during formation reactions. To illustrate this consider the fabrication of heavy metal fluoride glasses for potential optical fibre use, as further noted in Chapter 11. In order to gain some insight into the conditions occurring during reaction, a small amount of vanadium was incorporated into a glass composed mainly of ZrF_4, BaF_2 and NaF. When the glass was made in a nitrogen atmosphere, the colour was yellow-green and spectral analysis showed it to contain V^{3+} in octahedral sites. When a partial pressure of oxygen of about 0.1 atm. was introduced into the nitrogen, V^{5+} formed, which is colourless and so the glass loses its yellow-green hue. Surprisingly there was no trace found of the stable ion V^{4+} under any processing condition. This information allowed the reaction mechanisms occurring within the glass during fabrication to be determined, a task of considerable difficulty by other methods.

These examples show that the colour of a transition metal ion and its careful measurement using absorption spectra can give useful structural results in a variety of situations and often over a range of temperatures which remain inaccessible to other experimental techniques.

7.10 COLOURS FROM LANTHANIDE (RARE EARTH) IONS

The *lanthanides* (also called the *rare earths*) have electrons in partly filled 4f orbitals. The electron configurations of these important metals are given in Appendix 7.3. Many lanthanides show colours due to electron transitions involving the 4f-orbitals. However, there is a considerable difference between the lanthanides and the 3d transition metal ions. The 4f electrons in the lanthanides are well shielded beneath an outer electron configuration

Table 7.6 The colours characteristic of some lanthanide ions

Ion	f electron configuration	Characteristic colour
Ce^{3+}	$4f^1$	Yellow
Pr^{3+}	$4f^2$	Green
Nd^{3+}	$4f^3$	Lilac-violet
Pm^{3+}	$4f^4$	Pink
Sm^{3+}	$4f^5$	Pale yellow
Sm^{2+}	$4f^6$	Red/Green
Eu^{3+}	$4f^6$	Pink
Eu^{2+}	$4f^7$	Brown
Tb^{3+}	$4f^8$	Pink
Dy^{3+}	$4f^9$	Pale yellow
Dy^{2+}	$4f^{10}$	Brown
Ho^{3+}	$4f^{10}$	Yellow
Er^{3+}	$4f^{11}$	Pink
Tm^{3+}	$4f^{12}$	Green

($5s^2\ 5p^6\ 6s^2$) and so are little influenced by the crystal surroundings. Hence the important optical and magnetic properties attributed to the 4f electrons on any particular lanthanide ion are rather unvarying and do not depend significantly upon the host structure. These elements thus find use in phosphors (see Chapter 9) and lasers (see Chapter 13), where a host lattice can be chosen with respect to processing conditions without changing the desirable colour properties of the ion. Some typical lanthanide colours are listed in Table 7.6.

7.11 TRANSITION METAL AND LANTHANIDE PIGMENTS

Inorganic pigments are colorants used to enhance the appearance of an object. Pigments are incorporated as finely ground powders, and are often applied to surfaces as paints and inks. Although organic pigments (see Section 8.11) are usually brighter than inorganic pigments, they are not stable at moderate or high temperatures. This poses a major problem in the fabrication of decorative ceramics and glasses, as high temperatures are essential during manufacture. The use of transition metal (or more rarely lanthanide) compounds, which are added in small quantities to the batch, overcomes this difficulty. The colours generated are due to the d–d or f–f transitions described above.

Typical of simple oxides used in this context is green chromic oxide, Cr_2O_3. However, many more complex formulations are used. Cobalt aluminate spinel, $CoAl_2O_4$, which imparts a blue colour arising with the tetrahedrally coordinated Co^{2+} ions, and cobalt chromite spinel, $CoCr_2O_4$, which imparts a blue-green colour, arising from both the tetrahedrally coordinated Co^{2+} and octahedrally coordinated Cr^{3+} ions, are widely used pigments. Similarly, cobalt silicate, Co_2SiO_4, with the olivine structure, is a blue pigment, which again relies upon tetrahedrally coordinated Co^{2+} ions as the colour producer, and calcium chromium silicate, $Ca_3Cr_2Si_3O_{12}$, with the garnet structure, shows a green colour attributed to the octahedrally coordinated Cr^{3+} ions.

The use of such colorants is hardly new. Several thousand years ago the Egyptians synthesised blue objects using a colorant now known as "Egyptian Blue" and the Chinese synthesised both blue and purple artefacts using "Han Blue" and "Han Purple". These have recently been shown to be complex copper silicates. The formulae are $CaCuSi_4O_{10}$ for Egyptian Blue, $BaCuSi_4O_{10}$ for Han Blue and $BaCuSi_2O_6$ for Han Purple. The compounds themselves are ring silicates in which the colour is derived from crystal field splitting of the Cu^{2+} d-orbitals in a square planar environment. Bearing in mind the fact that the alkaline earth–copper–silicon–oxygen systems are complex and contain a bewildering variety of both coloured and non-coloured crystalline and glassy phases, the technological expertise of the craftsmen was considerable.

Enamels have also been produced for centuries by similar skilled craftsmen. Enamels are hard vitreous decorative coatings applied to metals by fusion of a base glass containing oxide pigments. Plate 7.5 shows and enamel lid on a trinket box.

The desire for bright colours in the past lead to the use of pigments which were certainly dangerous and which would not be allowed today. For example, Scheele's green, a bright green compound precipitated from "arsenious acid" with copper sulphate solution, which has been assigned an approximate formula $HCuAsO_3$, and Paris green, a mixed copper arsenic acetate of approximate formula $3Cu(AsO_2)_2.Cu(CH_3COO)_2$, were both widely used to colour much sought after green wallpaper in the 19th century. These compounds, however, are rather unstable and release toxic arsenic containing vapours in moist air. Indeed, the death of Napoleon, on the island of St. Helena, in 1821, is attributed to arsenic poisoning arising from the decomposition of Paris green pigments in the wallpapers of his accommodation.

7.12 ANSWERS TO INTRODUCTORY QUESTIONS

How can "Neon" signs be made to show a variety of colours?
The term "neon sign" has become rather general. Neon signs show different colours because a variety of gases other than neon are used. Each different gas will give a different colour.

How do sodium street lamps produce yellow light?

Sodium lamps produce light by exiting sodium atoms into higher energy levels by way of collisions with energetic electrons and the excited atoms and ions. These excited states all lose energy to regain the lowest energy ground state. In so doing they pass through a pair of energy levels at an energy above the ground state equivalent to that of yellow light. Thus, on returning to the ground state each excited sodium atom emits a photon of yellow light.

Why are transition metal compounds coloured?

Transition metal ions are not coloured when they are in the gaseous state. However, when they are incorporated into molecules or crystals the d-electron orbitals interact with the surrounding atoms and split into higher and lower energy groups. Colour arises due to transitions between these new energy levels. The energy level splitting is called crystal field or ligand field splitting and the transitions are called d–d transitions.

Why is ruby red and copper sulphate blue?

Ruby is crystalline Al_2O_3 (corundum), containing about 0.5% Cr^{3+}. The red colour arises from the crystal field splitting of the d-orbitals on the Cr^{3+} ions in a slightly distorted octahedral crystal field. Electron transitions from the lower energy levels to the upper absorb in the yellow and blue. The colour transmitted or reflected by ruby is a subtraction colour, the well known red with slight blue overtones.

Copper sulphate in its blue form is $CuSO_4.5H_2O$. The crystal structure shows that Cu^{2+} ions are surrounded by six oxygen ions in a distorted octahedral coordination. As in the case of Cr^{3+} the colour arises from electron transitions between the split 3d-orbitals in the octahedral crystal field. In this case the transitions are between the filled lower t_{2g} orbitals and the upper e_g orbitals. The absorption peak is in the near infrared, but it is broad enough to spill into the red end of the spectrum, endowing the crystals and solutions with the well known blue colour.

7.13 FURTHER READING

The origin of the colours in gases and transition metal ions is to be found in:

K. Nassau, The Physics and Chemistry of Color, Wiley-Interscience, New York (1983), Ch. 3 (gases), Ch. 5 (transition metals).

A simple and inexpensive way of observing line spectra from street lamps using plastic diffraction gratings is described by:

J. Walker, Scientific American, **250**, January 1984, p. 112.

The colours of gemstones are discussed by:

R. G. Burns, Mineralogical Applications of Crystal Field Theory, Cambridge University Press, Cambridge (1970).

The use of transition metal cations as structural probes in spinels and glasses is reported in:

F. S. Stone and R. J. D. Tilley, in Reactivity of Solids, ed. G. M. Schwab, Elsevier, Amsterdam (1965), p. 583.
O. Schmitz-Dumont, Bull. Chim. Soc. Fr., 1965, p. 1099.
M. D. Ingram and J. A. Duffy, J. Amer. Ceram. Soc., **53**, 1970, p. 317.
W. H. Turner and J. A. Turner, J. Amer. Ceram. Soc., **55**, 1972, p. 201.
P. W. France, S. F. Carter and J. M. Parker, Physics and Chemistry of Glasses, **27**, 1986, p. 32.

The defect chemistry which is involved in doping is explained in:

R. J. D. Tilley, Principles and Applications of Chemical Defects, Stanley Thornes, Cheltenham (1998), Ch. 7.

The electron configuration of atoms and ions is explained clearly in:

P. W. Atkins and L. Jones, Chemistry, 3rd edition, W. H. Freeman, New York (1997), Ch. 7.

The pigments Egyptian Blue, Han Blue and Han Purple are described in:

La couleur dans la peinture et l'emaillage de l'Egypte ancienne, eds S. Colinart and M. Menu, Edpuglia, Bari (1998). The paper by H. G. Wiedemann, G. Bayer and A. Reller, p. 195, is of especial relevance.

7.14 Problems and Exercises

1. Could you use a flame test to determine whether you had a mixture of lithium and sodium present in a sample?

2. Would you expect Fraunhofer lines to be present in spectra from the moon and planets?

3. What colours are emitted by a mercury lamp? How would the appearance of the light change at higher operating voltages and temperatures?

4. What is the main difference between the spectrum from a transition metal ion in a vacuum and one in a crystal?

5. How would you expect the crystal field splitting of an ion in a cubic (8-fold) site to compare with those in an octahedral or tetrahedral site?

6. Al_2O_3 is colourless, ruby (Al_2O_3 containing up to 1% Cr^{3+}) is red and Cr_2O_3 is green. Explain the colours of these three materials. (The unit cell of Al_2O_3 is hexagonal with a = 0.4763 nm, c = 1.3003 nm and that of Cr_2O_3 is hexagonal with a = 0.4960 nm, c = 1.3599 nm.)

7. The absorption spectrum of small amounts of Cu^{2+} impurity in MgO shows a band with a maximum at a wavelength of 900 nm and in ZnO at 1330 nm. When small amounts of Cu^{2+} are introduced into the spinel $MgAl_2O_4$ two bands are obtained, with maxima at 1660 nm and 710 nm. What structural conclusions can you draw from these results.

8. Silica gel, a material which gains and loses water easily, is used as a desiccant. For this purpose it contains a little Co^{2+}. When the material is dry it is blue and when it has absorbed water it is a pale pink. What causes these changes?

9. Nd^{3+} can be doped into a large number of host lattices including glass, garnets and the oxide $CaWO_4$. Although these all have quite different structures the colour of the compounds are all a similar lilac. Why is this?

10. Why is it not possible to use lanthanide ions as structural probes?

APPENDIX 7.1 THE ELECTRONIC CONFIGURATION OF ATOMS AND IONS

The electronic configuration of an atom or an ion describes the arrangement of the electrons around the nucleus. Each electron is assigned a set of four unique quantum numbers which correspond to the atomic orbital that the electron occupies. Each atom has a different number of electrons assigned to the various atomic orbitals available. The atomic orbitals form a set of shells which are filled from the lowest energy upwards.

The lowest energy shell is characterised by a principal quantum number 1 and contains only one atomic orbital called an *s-orbital*. This, like any atomic orbital, can contain either one or two electrons. The two atoms that these two alternatives correspond to are hydrogen (H) and helium (He). The electron configurations of these two atoms are written:

$$H \quad 1s^1$$
$$He \quad 1s^2$$

where the principal quantum number (1) is written first, the orbital (s) follows and then the number of electrons in the orbital as a superscript.

The next lowest energy shell is characterised by a principal quantum number 2 and contains one s-orbital and three *p-orbitals*, p_x p_y and p_z, all of which have the same energy. The s-orbital can contain up to two electrons, as above, and the three p-orbitals can contain a maximum of six electrons. The electron configurations of the atoms which make up the second shell, from lowest to highest energy, are:

Li	$1s^2 \, 2s^1$	or	$[He] \, 2s^1$
Be	$1s^2 \, 2s^2$	or	$[He] \, 2s^2$
B	$1s^2 \, 2s^2 \, 2p^1$	or	$[He] \, 2s^2 \, 2p^1$
C	$1s^2 \, 2s^2 \, 2p^2$	or	$[He] \, 2s^2 \, 2p^2$
N	$1s^2 \, 2s^2 \, 2p^3$	or	$[He] \, 2s^2 \, 2p^3$
O	$1s^2 \, 2s^2 \, 2p^4$	or	$[He] \, 2s^2 \, 2p^4$
F	$1s^2 \, 2s^2 \, 2p^5$	or	$[He] \, 2s^2 \, 2p^5$
Ne	$1s^2 \, 2s^2 \, 2p^6$	or	$[He] \, 2s^2 \, 2p^6$

The second shell is now full. Note that in order to write the configuration in a compact form the inner filled shell is represented by the symbol of the atom with that configuration, He in this case.

The next energy shell is characterised by a principal quantum number 3 and contains one s-orbital, three p-orbitals p_x, p_y and p_z, and five *d-orbitals*, d_{xy}, d_{xz}, d_{yz}, $d_{x^2-y^2}$ and d_{z^2}. The s-orbital can contain up to two electrons and the three p-orbitals can contain a maximum of six electrons, as before. The five d-orbitals can contain up to 10 electrons. Atoms with partly filled d orbitals are called transition metals. The electron configurations of the atoms which make up the third shell, from lowest to highest energy, are:

Na	$1s^2 \, 2s^2 \, 2p^6 \, 3s^1$	or	$[Ne] \, 3s^1$
Mg	$1s^2 \, 2s^2 \, 2p^6 \, 3s^2$	or	$[Ne] \, 3s^2$
Al	$1s^2 \, 2s^2 \, 2p^6 \, 3s^2 \, 3p^1$	or	$[Ne] \, 3s^2 \, 3p^1$
Si	$1s^2 \, 2s^2 \, 2p^6 \, 3s^2 \, 3p^2$	or	$[Ne] \, 3s^2 \, 3p^2$
P	$1s^2 \, 2s^2 \, 2p^6 \, 3s^2 \, 3p^3$	or	$[Ne] \, 3s^2 \, 3p^3$
S	$1s^2 \, 2s^2 \, 2p^6 \, 3s^2 \, 3p^4$	or	$[Ne] \, 3s^2 \, 3p^4$
Cl	$1s^2 \, 2s^2 \, 2p^6 \, 3s^2 \, 3p^5$	or	$[Ne] \, 3s^2 \, 3p^5$
Ar	$1s^2 \, 2s^2 \, 2p^6 \, 3s^2 \, 3p^6$	or	$[Ne] \, 3s^2 \, 3p^6$

The energy of the 4s-orbital is close to that of the 3d-orbitals and is usually filled before the 3d group. The electron configuration of the 3d transition metals is given in Appendix 7.2.

The filling of the 4th, 5th and subsequent shells follows along the same lines as above. In heavier atoms there is often some uncertainty in the order in which the orbitals are filled. This will be observed for example in some of the atoms listed in Appendix 7.3. The completely filled $ns^2 np^6$ configurations (which correspond to the inert gases) used to write the electron configurations in a compact form, are:

$$
\begin{array}{ll}
\text{He} & 1s^2 \\
\text{Ne} & \text{[He]} \ 2s^2 \ 2p^6 \\
\text{Ar} & \text{[Ne]} \ 3s^2 \ 3p^6 \\
\text{Kr} & \text{[Ar]} \ 3d^{10} \ 4s^2 \ 4p^6 \\
\text{Xe} & \text{[Kr]} \ 4d^{10} \ 5s^2 \ 5p^6
\end{array}
$$

The electron configuration of ions is written in an identical fashion. Cationic configurations can usually be derived from that of the parent atoms by removing a small number of electrons from the atomic orbitals last filled and anionic configurations by adding electrons to these same orbitals.

APPENDIX 7.2 THE 3d TRANSITION METALS

The ten 3d transition metal elements are found in Period 4 of the periodic table:
K, Ca, **Sc, Ti, V, Cr, Mn, Fe, Co, Ni, Cu, Zn**, Ga, Ge, As, Se, Br, Kr. They are characterised by having partly filled 3d atomic orbitals. Cuprous, Cu^+, and zinc, Zn^{2+}, ions do not behave as transition metal ions as the 3d-orbitals are completely filled.

Name	Symbol	Electronic configuration of atom	Ion	d electron configuration of ion
Scandium	Sc	[Ar] $3d^1 4s^2$	Sc^{3+}	d^0
Titanium	Ti	[Ar] $3d^2 4s^2$	Ti^{4+}	d^0
			Ti^{3+}	d^1
			Ti^{2+}	d^2
Vanadium	V	[Ar] $3d^3 4s^2$	V^{5+}	d^0
			V^{4+}	d^1
			V^{3+}	d^2
			V^{2+}	d^3
Chromium	Cr	[Ar] $3d^5 4s^1$	Cr^{3+}	d^3
Manganese	Mn	[Ar] $3d^5 4s^2$	Mn^{4+}	d^3
			Mn^{3+}	d^4
			Mn^{2+}	d^5
Iron	Fe	[Ar] $3d^6 4s^2$	Fe^{3+}	d^5
			Fe^{2+}	d^6
Cobalt	Co	[Ar] $3d^7 4s^2$	Co^{4+}	d^5
			Co^{3+}	d^6
			Co^{2+}	d^7

Name	Symbol	Electronic configuration of atom	Ion	d electron configuration of ion
Nickel	Ni	[Ar] $3d^8 4s^2$	Ni^{2+}	d^8
Copper	Cu	[Ar] $3d^{10} 4s^1$	Cu^{2+}	d^9
			Cu^+	d^{10}
Zinc	Zn	[Ar] $3d^{10} 4s^2$	Zn^{2+}	d^{10}

APPENDIX 7.3 THE LANTHANIDE OR RARE EARTH ELEMENTS

The 14 lanthanide or rare earth elements are found in Period 6 of the periodic table:

Cs, Ba, **(La)**, **Ce, Pr, Nd, Pm, Sm, Eu, Gd, Tb, Dy, Ho, Er, Tm, Yb, (Lu)**, Hf, Ta, W, . . .

The electron configuration of some of these atoms is uncertain and neither lanthanum nor lutetium behave as typical lanthanides although both are frequently included in the group.

Name	Symbol	Electronic configuration of atom[a]	Ion	f electron configuration of ion
Lanthanum	La	[Xe] $5d^1 6s^2$ or [Xe] $4f^1 6s^2$	La^{3+}	$4f^0$
Cerium	Ce	[Xe] $4f^1 5d^1 6s^2$ or [Xe] $4f^2 6s^2$	Ce^{4+}	$4f^0$
			Ce^{3+}	$4f^1$
Praseodymium	Pr	[Xe] $4f^3 6s^2$	Pr^{4+}	$4f^1$
			Pr^{3+}	$4f^2$
Neodymium	Nd	[Xe] $4f^4 6s^2$	Nd^{3+}	$4f^3$
Promethium	Pm	[Xe] $4f^5 6s^2$	Pm^{3+}	$4f^4$
Samarium	Sm	[Xe] $4f^6 6s^2$	Sm^{3+}	$4f^5$
			Sm^{2+}	$4f^6$
Europium	Eu	[Xe] $4f^7 6s^2$	Eu^{3+}	$4f^6$
			Eu^{2+}	$4f^7$
Gadolinium	Gd	[Xe] $4f^7 5d^1 6s^2$	Gd^{3+}	$4f^7$
Terbium	Tb	[Xe] $4f^9 6s^2$	Tb^{4+}	$4f^7$
			Tb^{3+}	$4f^8$
Dysprosium	Dy	[Xe] $4f^{10} 6s^2$	Dy^{3+}	$4f^9$
Holmium	Ho	[Xe] $4f^{11} 6s^2$	Ho^{3+}	$4f^{10}$
Erbium	Er	[Xe] $4f^{12} 6s^2$	Er^{3+}	$4f^{11}$
Thulium	Tm	[Xe] $4f^{13} 6s^2$	Tm^{3+}	$4f^{12}$
Ytterbium	Yb	[Xe] $4f^{14} 6s^2$	Yb^{3+}	$4f^{13}$
			Yb^{2+}	$4f^{14}$
Lutetium	Lu	[Xe] $4f^{14} 5d^1 6s^2$	Lu^{3+}	$4f^{14}$

[a] [Xe] = $1s^2\, 2s^2\, 2p^6\, 3s^2\, 3p^6\, 3d^{10}\, 4s^2\, 4p^6\, 4d^{10}\, 5s^2\, 5p^6$.

APPENDIX 7.4 ATOMIC AND IONIC ENERGY LEVELS

The electron configurations of atoms (or ions) given do not take electron–electron repulsion into account in a very comprehensive way. In fact, they are worked out by assuming that we only have one electron circling a nucleus surrounded by negative cloud made up of all of the other electrons present. Because only one electron is involved in the computations the quantum numbers are called hydrogen-like or one-electron symbols and they are given lower case letter labels. This means that the electron configurations given provide only a very approximate idea of the energy state of the atom or ion.

Atomic or ionic spectra, on the other hand, consist of a series of lines which give information on the exact energy difference between two energy levels in the species under investigation. Atomic spectra have thus allowed the exact energy levels of atoms to be constructed. The energy levels of an isolated atom are given labels called term symbols. A term is a set of states or energy levels which are very similar in energy. Transitions between these terms gives rise to the observed line spectrum of an atom. Each term is written as:

$$\text{term} = {}^{2S+1}L_J$$

where L is a many-electron quantum number describing the total orbital angular momentum of all of the electrons surrounding the atomic nucleus, S is a many-electron quantum number representing the total electron spin and J is a many-electron quantum number representing the total angular momentum of the electrons. The superscript (2S+1) is called the *multiplicity* of the term and gives the number of levels present, which is equal to the number of values of J present. Upper case letters are used to make it clear that all electrons are included and to differentiate them from the hydrogen-like configurations. Terms therefore apply to the overall energy state of the atom or ion as a whole.

The quantum numbers L are not given numerical values, but letter symbols corresponding to those for the hydrogen-like configurations. Thus L = 0, 1, 2, 3 . . . are written as S, P, D, G . . . . A typical term would then be written $^3P_2$. The multiplicity of 3 tells us that there are three energy levels associated with this term, and when written out fully these would be $^3P_0$, $^3P_1$ and $^3P_2$.

It is not especially easy to go from atomic configurations to term symbols and associated energies. However, this task has been completed for most atomic and ionic species of interest and when this information is required we have only to look up the relevant tables of data.

The nomenclature just described is not adequate to describe either molecular energy levels or the energy levels of atoms in crystal fields. In these cases a terminology based upon symmetry is most often encountered.

CHAPTER 8

COLOUR FROM MOLECULES

Why is deep water tinted blue?
What colours roses red and cornflowers blue?
Why does the colour of red wine change with age?

8.1 THE SPECTRA AND ENERGY LEVELS OF MOLECULES

Whereas a gas of atoms emits light at precise wavelengths to give a series of sharp lines, molecules emit a band of wavelengths. Thus the line spectrum of an atom will be broadened into a band spectrum from a molecule. Each band generally has one sharp side and a diffuse gradually fading side to it. Under high resolution the bands are resolved into closely spaced series of lines. The origin of the bands lies in the addition of new energy levels due to vibration and rotation of the molecule to each electronic energy level, as indicated schematically in Figure 8.1. Thus each single line corresponding to an electronic transition in a single atom can become a closely spaced set of lines (a band) in a molecule.

The electronic energy levels are separated by energies of the order of 6×10^{-19} J (3.7 eV), and these produce spectral lines in the ultraviolet region. The energy increments of the vibrational levels is about a tenth of that between the electronic energy levels, i.e. 6×10^{-20} J (0.37 eV) and gives rise to absorption and emission at infrared wavelengths. The energy steps of the rotational levels are about a hundred times smaller again, at approximately 6×10^{-22} J (0.0037 eV), giving rise to microwave absorption and emission.

Colours due to molecular (and some atomic) transitions are seen in the upper atmosphere as the spectacular displays *aurora borealis* or *aurora australis*. The auroras form at heights of 100 to 1000 km above the polar regions. Very energetic particles, mostly electrons together with some protons, mainly originating in the sun, spiral along the Earth's magnetic field lines towards the poles. When they reach the tenuous outer limits of the atmosphere they collide with and excite the atoms and molecules encountered. These excited species lose energy by radiating in part in the visible and give rise to the remarkable shifting curtains of colour seen in the polar regions.

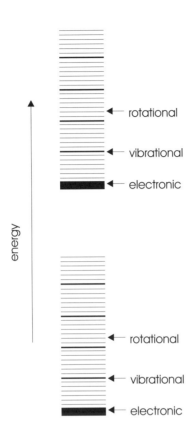

Figure 8.1 Schematic diagram of the electronic, vibrational and rotational energy levels of a molecule (not to scale). Each electronic energy level has additional associated energy levels due to molecular vibration and rotation. The electronic energy levels are separated by approximately 6×10^{-19} J (3.7 eV); the vibrational levels by approximately 6×10^{-20} J (0.37 eV) and the rotational levels by approximately 6×10^{-22} J (0.0037 eV)

The major components contributing to the colours are nitrogen molecules, N_2, and oxygen atoms, O. Nitrogen molecules can become ionised to N_2^+ which then can re-capture an electron to leave an excited nitrogen molecule, N_2^*. This species then decays to the ground state, giving out light in the process.

$$N_2 + e^- \rightarrow N_2^+ + 2e^-$$
$$N_2^+ + e^- \rightarrow N_2^* + \text{violet and blue light}$$
$$N_2^* \rightarrow N_2 + \text{pink light}$$

Oxygen atoms, O, which are more common than oxygen molecules in the outer regions of the atmosphere, are formed by photodissociation of O_2 under intense ultraviolet irradiation in these near space conditions. These are excited by electron bombardment to form O^* species which return to the ground state by the emission of whitish-green and crimson light.

$$O + e^- \rightarrow O^* + e^- \text{ (with a lower energy)}$$
$$O^* \rightarrow O + \text{whitish green and crimson light}$$

Colour from molecular transitions is also seen in the blue region around a candle flame, sketched in Figure 8.2a. The blue colour is mainly produced by excitations of two molecular fragments formed in the flame, C_2 and CH. The strongest CH band is at 432 nm in the indigo region of the spectrum, while C_2 has a strong band in the green with less intense bands in the blue and violet regions. The main bands are represented schematically in Figure 8.2b.

The energy involved in transitions between the rotational and vibrational energy levels is small and gives rise to infrared and microwave absorption and so are not usually perceived directly. An exception occurs in the water molecule. This is an angular molecule with a *bending mode* of vibration which absorbs energy in the infrared, at a wavelength of 6273 nm. In addition, two *stretching modes*, in which the bonds in the molecule lengthen and shorten, also occur. One of these, in which the bonds lengthen and shorten together, the *symmetrical mode*, absorbs energy at 2730 nm. The other, in which one bond lengthens as the other shortens, the *anti-symmetrical mode*, absorbs energy at 2662 nm. These absorption wavelengths are far from the visible, and as we know, water molecules in the atmosphere are colourless. Indeed, before the advent of Rayleigh's scattering theory it was believed that the blue of the sky could be due to water in the atmosphere, a supposition which is not correct.

Although the three absorption peaks for molecules of water are far from the visible, they can combine to produce *overtones*, which are harmonics, and *combinatorial tones*, which are sums, of the fundamental frequencies. For example, if we set the frequencies of the absorption maxima as ν_1, ν_2, and ν_3, the overtones are of the form $2\nu_1$ and the combinatorial tones are of the form $2\nu_2 + \nu_3$. The existence of these terms extends the spectrum of water molecules much closer to the visible but still not close enough to present any sensation of colour to the eye. However these terms do result in a rather complex absorption spectrum for water molecules.

There are two important ways in which infrared absorption bands can be moved to shorter wavelengths. The mass of the atoms in the bonds can

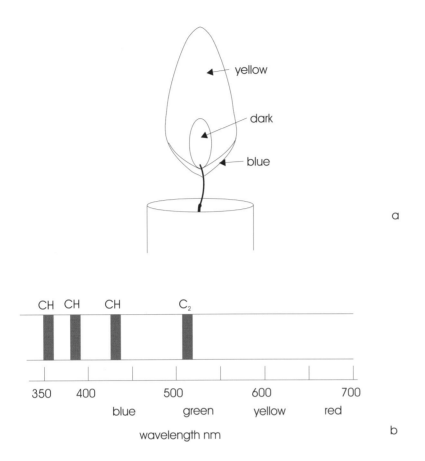

Figure 8.2 (a) The blue outer sheath at the base of a candle flame arises from CH and C_2 molecules. (b) The spectrum, shown schematically, indicates the positions of the main emission bands from these species

be increased and the bonds can be made stronger. (Both of these attributes are of importance in the improved optical transmission of heavy-metal fluoride glass optical fibres with respect to silica fibres, discussed in Chapter 11.) In the case of water, the atoms are of fixed mass, but in the liquid and solid states, the bonding is altered. The change comes about by the addition of a new component, hydrogen bonding, which links the molecules together by additional liaisons. Although hydrogen bonding is weak, with a bond energy of approximately 20 kJ mol^{-1} compared to the H–O bond strength of 463 kJ mol^{-1}, it is of enormous significance and, amongst other things, controls the conformation of many of the important molecules of life. In the case of water, the change in bonding is enough to shift the absorption spectrum to lower wavelengths. Although the shift is relatively small it is enough to move the short wavelength edge into the visible. At this stage the strength of the absorption is low, but it is sufficient to remove a small fraction of red and orange. This is adequate to give sizeable bodies of pure water or ice a pale (watery!) blue colour.

8.2 MOLECULAR ORBITALS

As well as the additional energy levels due to vibration and rotation that appear in the energy levels of molecules, other terms can also arise which change the electronic energy levels considerably. The most important of these is due to the chemical bonds which maintain the coherence of the molecule. At the simplest level chemical bonds can be thought of as localised between pairs of atoms and formed by overlap of two atomic orbitals, one on each atom of the pair. In a more accurate representation, the overlapping atomic orbitals can be considered to extend over some or all of the atoms in the molecule to form *molecular orbitals*. Electronic transitions between molecular orbitals occur at different energies to those between the electronic energy levels of the constituent atoms of the molecule and are responsible for many of the colours observed in plants and animals. These colours have been rationalised in terms of the molecular orbital theory of molecules.

The situation is summarised in Figure 8.3a,b. For the intensely coloured organic molecules that are of most interest here, the highest molecular orbital containing electrons is usually a π molecular orbital, as indicated in Figure 8.3a. The orbital is said to be a bonding orbital because its energy relative to other energy levels in the molecule allows the electrons within it to contribute to the chemical bonds between the atoms of the molecule. In a shorthand notation this orbital is often referred to as the *highest occupied molecular orbital*, or HOMO. The orbital with the next highest energy, which will not contain any electrons, is known as an antibonding orbital and carries the label π^*. Such an orbital is called the *lowest unoccupied molecular orbital* or LUMO for short. Between the π and π^* orbitals one or more *non-bonding orbitals*, n, are sometimes present, as drawn in Figure 8.3b. These are orbitals which contain electrons which do not contribute to bonding in the molecule and might be, for example, a lone pair of electrons associated with a nitrogen atom. An occupied n orbital might therefore be the HOMO in the system, as in Figure 8.3b.

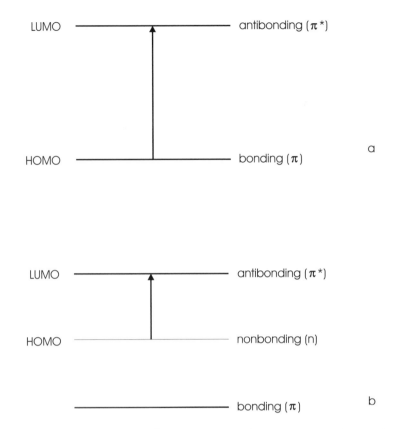

Figure 8.3 Schematic diagram of the energy levels of importance in producing colour in many organic compounds. (a) The major colour producing electronic transition, π to π^*, arrowed, is from the highest occupied π molecular orbital (HOMO) to the lowest unoccupied π^* molecular orbital (LUMO). (b) When a non-bonding orbital is present a colour producing electronic transition, n to π, arrowed, can be from the highest occupied n molecular orbital (HOMO) to the lowest unoccupied π^* molecular orbital (LUMO)

The most important electronic energy transition for colour production is electron transfer from a π-type HOMO to the π^*-type LUMO. This type of transition is referred to as a π to π^* transition. Such transitions give rise to intense absorption bands with high absorption coefficients and are found in molecules containing conjugated single and double bonds, described in Section 8.4, below. Transitions from an n-type HOMO to the π^* LUMO, n to π^* transitions, are also possible. These give rise to less intense absorption bands than the π to π^* transitions but are none the less important. They occur in molecules containing a =C=O group (the ketones) and are the source of colour in a variety of dyes.

8.3 CHROMOPHORES, CHROMOGENS AND AUXOCHROMES

The earliest studies in organic chemistry showed that colours in organic molecules could be manipulated experimentally. For example, it was found that coloured organic materials were turned colourless by the addition of hydrogen and returned to their original colours by the removal of hydrogen. To try to rationalise the experimental observations the German chemist Witt suggested, in 1876, long before quantum theory and X-ray structural studies, a series of guidelines relating to the colour of organic molecules. The source of the colour in a molecule was supposed to be one or more "colour bearing" small groups of atoms with multiple bond configurations, called *chromophores*. (This terminology is not restricted to organic molecules and is often used in inorganic chemistry to denote a group of atoms or ions which cause colour.) Some important chromophores are listed in Table 8.1. The compound containing the chromophore(s) was called a *chromogen*. The depth of colour of the chromogen was proportional to the number of chromophores present. It was recognised that some groups in an organic molecule, called *auxochromes*, also played a role. Although auxochromes did not produce colour themselves they had the effect of intensifying the colour of a molecule if a chromophore was present. The important auxochromes are hydroxyl (–OH), keto groups (=C=O) and groups of atoms containing nitrogen.

Theoretical calculations show that the presence of chromophores decreases the energy between the HOMO and the LUMO. The more chromophores in a molecule the greater the decrease in energy is found to be. Thus, in cases where the main absorption band of a parent molecule lies in the ultraviolet, the absorption band of a daughter molecule containing one or more chromophores is moved towards the visible. In suitable cases the result is the transformation of a colourless parent compound into an intensely coloured daughter molecule.

Table 8.1 Some chromophores

Group	Formula	Group	Formula
Nitro	$-NO_2$	Azoxy	$-N=N-O-$
Carbonyl	$=CO$	Nitroso	$-NO$
Azo	$-N=N-$	Azoamine	$-N=N-NH-$
Thiocarbonyl	$=CS$	Ene	$>C=C<$

8.4   CONJUGATED MOLECULES

Although the >C=C< *double bond* arrangement linking two carbon atoms was regarded as a chromophore, an isolated >C=C< group has a π to π^* absorption band centred at a wavelength near to 160 nm in the far ultraviolet and so does not lead to colour in a molecule. In agreement with the concept of chromophores just described, a dramatic change occurs when a number of these units are arranged in an alternating single-bond double-bond sequence, to form a series of *conjugated* double bonds. For example, whereas the absorption maximum of ethene ($CH_2=CH_2$) is at 162.5 nm, that of the compound $CH_3-CH=CH-CH=CH-CH=CH-CH=CH-CH_3$, or $CH_3-(CH=CH)_4-CH_3$, with four conjugated double bonds, has an absorption maximum at 296 nm. Colour is first found in the molecule containing six conjugated double bonds, $CH_3-(CH=CH)_6-CH_3$, in which the absorption maximum encroaches into the blue end of the spectrum, causing the molecule to appear yellow.

Two of the more important conjugated molecules are α- and β-carotene. These substances, when pure, form deep purple-red crystals. They are found in carrots and give these vegetables their orange colour. The structure of the α-compound is drawn in Figure 8.4a. Closely related to these are the pigments crocin and crocetin. These are an orange-yellow colour and are familiar as saffron, which is derived from crocus pollen. The structures of these compounds, shown in Figure 8.4c, d, clearly show the conjugated backbone of the molecules. (An examination of Figure 8.4 will show that the sequence of double and single bonds can be drawn in more than one way. This has consequences for colour generation which are outlined in Appendix 8.1.)

8.5   CONJUGATED BONDS CIRCLING METAL ATOMS

Systems of conjugated bonds which circle a metal atom can give rise to rich colours. Many molecules containing this arrangement are important to life processes. The two main classes of compounds in this context are called *porphyrins* and *phthalocyanines*. Of these, the porphyrin *chlorophyll*, the source of the green colour of plants, is surely among the most important molecules known.

Chlorophyll is found in four forms, called *chlorophyll a*, *chlorophyll b*, *chlorophyll c* and *chlorophyll d*. Higher plants and green algae contain chlorophyll a and b in a ratio of about 3:1. Red algae contain mainly

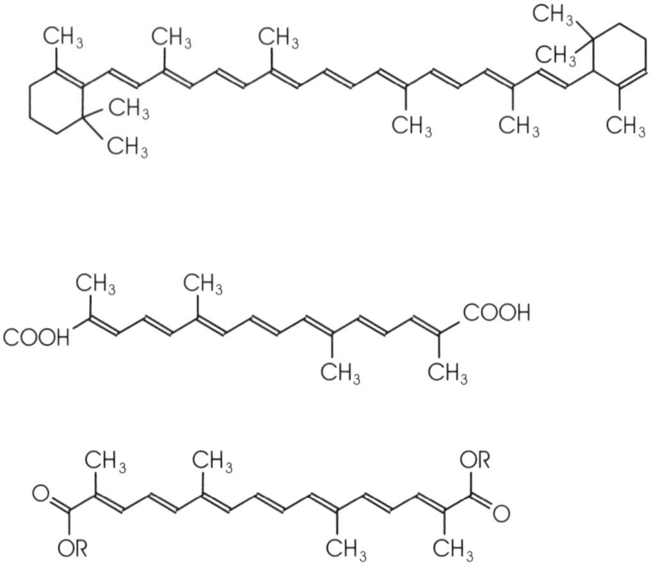

Figure 8.4 The structures of the conjugated molecules (a) α-carotene, purple-red, (b) crocetin, orange-yellow, and (c) crocin, orange-yellow. In these and succeeding figures carbon (C) and hydrogen (H) atoms are omitted from the main skeleton of the molecule and only the carbon–carbon single and double bonds are depicted, as single and double lines, respectively. At the periphery of the molecule the atoms are indicated. Apart from C and H, O represents oxygen and R symbolises a group of bonded atoms not important for colour production

chlorophyll a and some chlorophyll d. Chlorophyll c (together with some chlorophyll a) is found in many marine algae. The central core of all chlorophyll molecules is a magnesium atom surrounded by a sequence of alternating double and single bonds, as drawn in Figure 8.5. Chlorophyll molecules absorb strongly in the blue and red parts of the spectrum. The colour reflected by leaves corresponds to those wavelengths not strongly absorbed, which are the greens and to some extent the yellows. This provides an example of subtractive coloration.

Iron forms the core of another vital porphyrin molecule, *heme*, illustrated in Figure 8.6a. It forms the central feature of the molecule haemoglobin, which transports oxygen to and fro in the cells of the body. It is also responsible for the colour of blood, as the group has strong absorption maxima in the green-yellow part of the spectrum. Reds and violets are therefore reflected to produce the coloration of fresh blood. Interestingly, the mineral Fe_2O_3, called *hematite*, was so named because its red colour was reminiscent of fresh blood. In Fe_2O_3 the colour arises from transitions

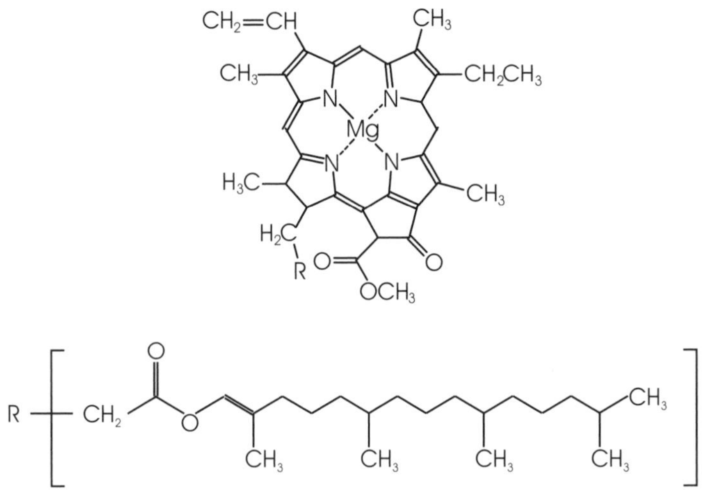

Figure 8.5 The structure of chlorophyll a. The molecule is built around a central magnesium (Mg) atom, linked to four nitrogen (N) atoms. This arrangement forms a typical porphyrin ring structure. The structure of the group R is drawn below the molecule

involving the 3d electron levels on the Fe^{3+} ions. In haemoglobin, despite the fact that an iron atom is present, the colour arises from π to π^* transitions and not from the iron at all!

The phthalocyanins, discovered early in the 20th century but not characterised structurally until the 1930s, are rather similar metal-centred molecules. The structure of copper phthalocyanine, a blue compound, is reproduced in Figure 8.6b. Although copper phthalocyanine is not found in nature, rather similar blue compounds do occur. They are found in the blue blood of hermit crabs and related crustaceans. The blue colour arises from π to π^* transitions in copper containing *hemocyanin* molecules which transport oxygen and play an analogous role to the haemoglobins in mammalian blood. Thus we find, as in the case of heme, that the colour of the molecule is similar to the crystal field colour of the central cation, but arises from a quite different mechanism.

8.6 THE DETECTION OF METAL IONS

The reactions of metallic cations with organic molecules to form complex molecules or *complexes*, rather like those just discussed, has been used to

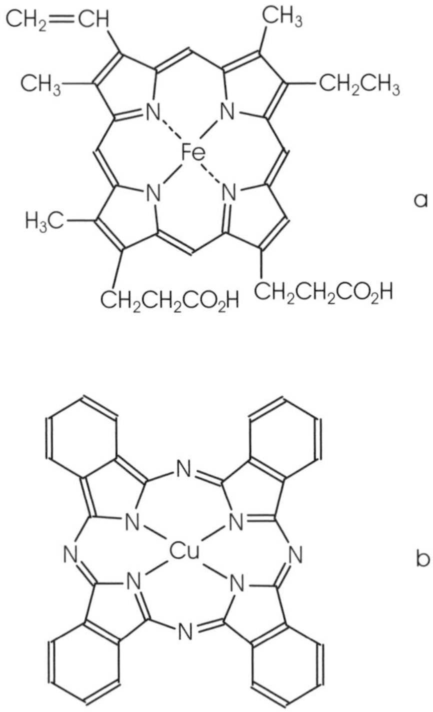

CH_2=CH CH_3

CH_3—

—CH_2CH_3

N   N

Fe

N   N

H_3C—

a

$CH_2CH_2CO_2H$ $CH_2CH_2CO_2H$

b

Figure 8.6 The structures of (a) red heme, a porphyrin, and (b) blue copper phthalocyanine, a phthalocyanine. The colours are produced within the organic structure and not by the transition metal cations

detect small quantities of metals in a solution. The reactions utilised are those in which the cation in question reacts to form a complex, the colour of which is indicative of the cation present. This technique is simple to apply and can readily give qualitative information on impurities at a parts per million level of discrimination. The procedure is simple. A drop of the solution to be tested is placed on a filter paper or into a well on a white ceramic plate. To this is added a drop of the necessary reagent and the colour produced, if any, is observed. Difficulties lie in ensuring that the solutions are free from contamination and that the pH is correct, as the colours seen are often pH dependent. If the amount of product is evaluated the method becomes quantitative.

An illustration of this technique is provided by the detection of nickel and palladium using the organic compound dimethyl glyoxime. In a basic solution this molecule will produce an intensely scarlet precipitate in the

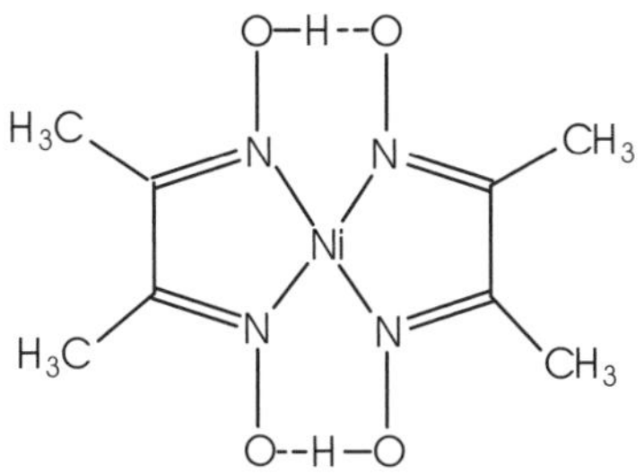

Figure 8.7 The structure of the intensely scarlet complex of nickel (Ni) with dimethylgloxime to form bis(dimethylgloximate)nickel II. The dotted lines represent hydrogen bonds

presence of Ni^{2+} cations. The structure of the complex is shown in Figure 8.7, which shows the central Ni cation to be surrounded by a square of four nitrogen atoms from coordination to two glyoxime molecules. The compound is called bis(dimethylglyoximate)nickel II. If the solution is acidified the scarlet colour will disappear. Should any palladium be present a yellow compound with a similar structure will form.

The number of organic molecules that can bind to metal ions to produce coloured products is enormous and so the majority of cations can be conclusively identified using this method. Details will be found in the Further Reading section at the end of this chapter.

8.7 NATURALLY OCCURRING COLORANTS

The *melanins* are responsible for most of the black and brown colours found in nature, including the brown colour of hair and skin and the brown colour which appears on damaged or cut fruit. They are a group of colorants whose structures, and the relationships between structure and colours are still poorly understood. *Eumelanin*, mainly responsible for blacks, is a complex polymer of o-diphenols or other related molecules. The likely structure of a fragment forming the building block of one of the many forms of eumelanin, revealing an extensive array of conjugated double bonds, is drawn in Figure 8.8. In the naturally occurring colorant many of these fragments are linked together to form the polymer. Many browns, red-browns and tans are attributed to the presence of another melanin variant, *phaeomelenin*. The structure of this material is less well understood than that of eumelanin and further studies in this area are needed before the nature of the various colour forms is clarified.

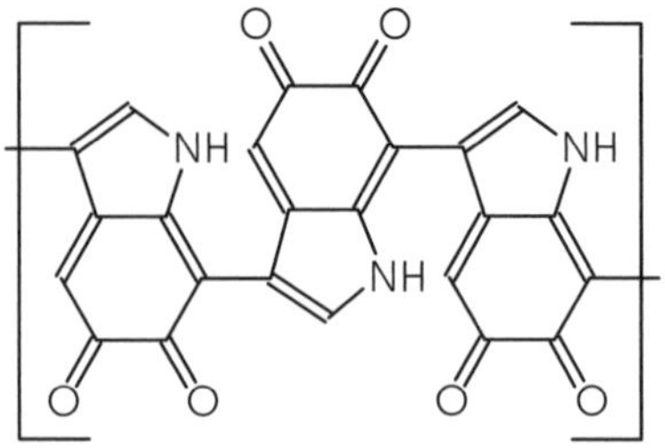

Figure 8.8 The likely structure of a fragment forming the building block of one of the many forms of the eumelanin polymer. In the naturally occurring colorant many of these fragments are linked together via the unterminated chemical bonds at the right and left periphery of the fragment

Flavonoids are a diverse and colourful group of pigments giving rise to whites, yellows, reds and blues. Structurally they are related to the compound *flavone*, shown in Figure 8.9a. This compound was first isolated from the primrose *Primula malacoides*. The influence of increasing the number of auxochrome (–OH) groups on colour is well illustrated in the sequence of flavonoid compounds *flavone* (Figure 8.9a), which is colourless, *flavonol* (Figure 8.9b), which is yellow and *quercetin* (Figure 8.9c), which is orange. Plate 8.1 shows a yellow water lily (*Nymphaea* hybrid, Chromatella) containing a flavone colorant and a brown butterfly (*Maniola jurtina*) mainly coloured by melanins.

Many of the blues and reds of flowers are derived from a group of flavonoid related compounds called *anthocyanidins*, all of which absorb strongly in the green region of the spectrum. The flowers therefore reflect reds and blues, which combine to produce a series of purple-red colours. Anthocyanidins have the general structure drawn in Figure 8.10a. The various anthocyanidin plant pigments found in nature differ from each other in having different groups, labelled R_1 and R_2, attached to the structure. Table 8.2 gives information for some of these and the flowers in which they occur. Anthocyanidins become *anthocyanins* when linked to sugars. These are also widely distributed plant colorants. For example, when linked to the sugar glucose, the anthocyanin colorant of chrysanthemums, chrysanthemumin, and asters, asterin, is formed.

The colour of these molecules is pH dependent, a feature used in nature by a number of plants. An example of the changes which can occur in the molecule cyanidin is given in Figure 8.10b–d. Such variations can be noticed in the colours of some flowers grown in soils of differing acidity. For example, hydrangeas grown on acid soil are blue while those grown on

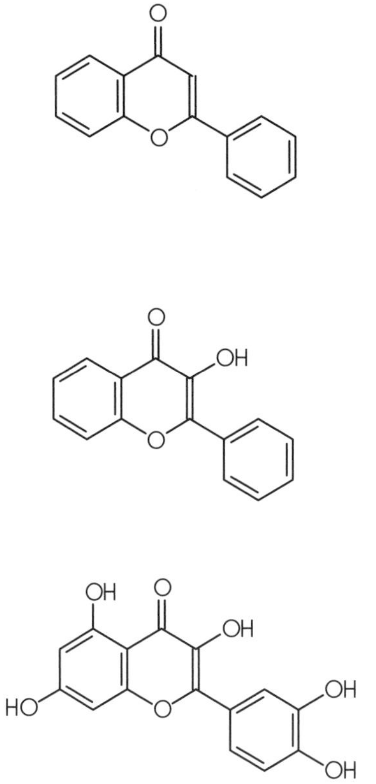

Figure 8.9   The structures of (a) flavone (colourless), (b) flavonol (yellow) and (c) quercetin (orange)

alkaline soil are pink. The red colour of poppies and many red roses is due to the flavonoid pigment cyanin. This same compound causes the blue colour in cornflowers and geraniums. These differences, brought about by changes in the pH of the sap within the plants, is illustrated in Plate 8.2a,b.

8.8   THE COLOUR OF RED WINE

The difference between the colour of red and white wines rests with the presence or absence of rather complex anthocyanin related materials. These are found in the outer layers of the skins of black grapes and are incorporated into the wine by allowing the skins to remain in contact with the pressed grape juice. The red colorants in red wine are a group of cations given the general name of *flavylium* cations. (Flavylium is a general term

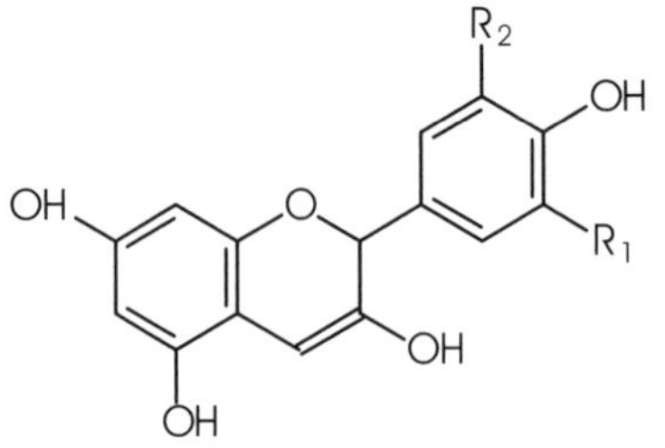

Figure 8.10 (a) The general structure of an anthocyanidin, where R_1 and R_2 represent groups which have an influence on the colour of the molecule. (b, c, d) Three colour forms of the molecule cyanidin: (b) red, in acid solution; (c) violet in neutral solution; (d) blue in alkaline solution

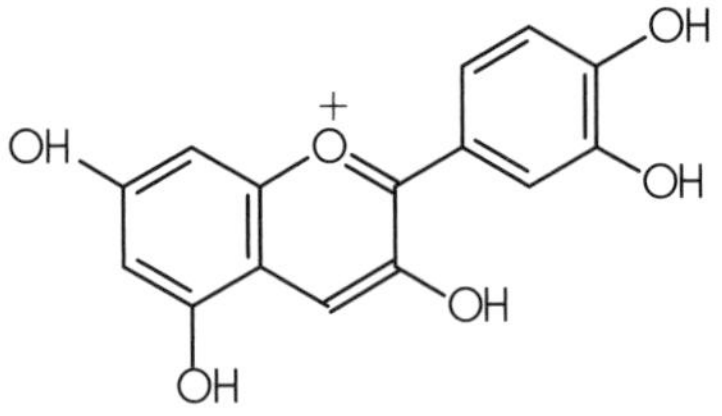

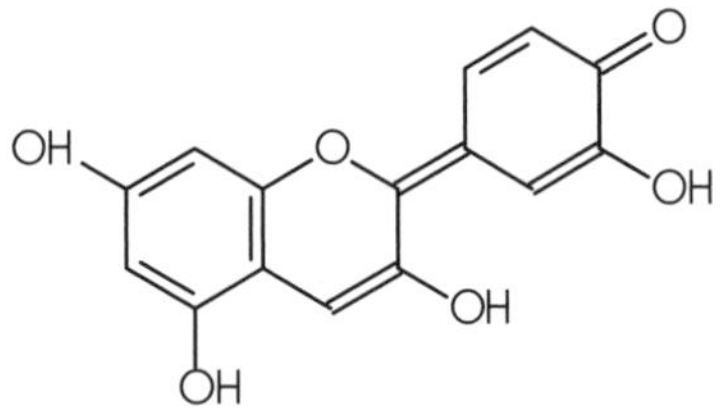

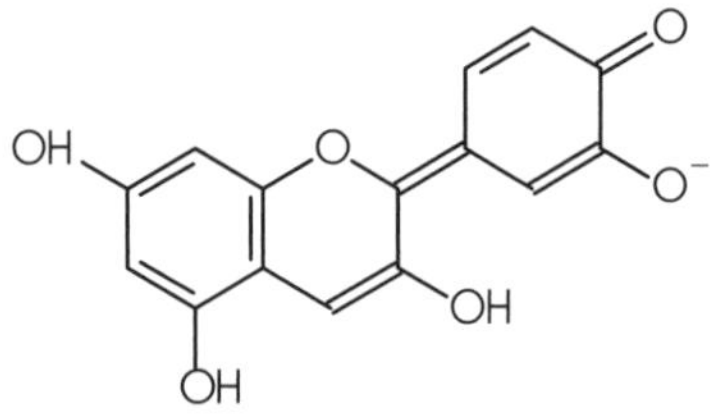

Table 8.2 Some anthocyanidins found in plant flowers

Anthocyanidin	Plant source	Typical colour[a]	R_1	R_2	Absorption maximum (nm)
Cyanidin			OH	H	535
Pelargonidin	Pelargoniums	Pink-red	H	H	520
Paeonidin	Paeonies	Red	$O.CH_3$	H	532
Delphidinin	Dephiniums	Blue	OH	OH	546
Petunidin	Petunias	Red	$O.CH_3$	OH	543
Malvidin	Mallows	Pink	$O.CH_3$	$O.CH_3$	542

[a] Horticulture has produced a vast range of colour types in all of these flower groups. Only the native colour is given in the table.

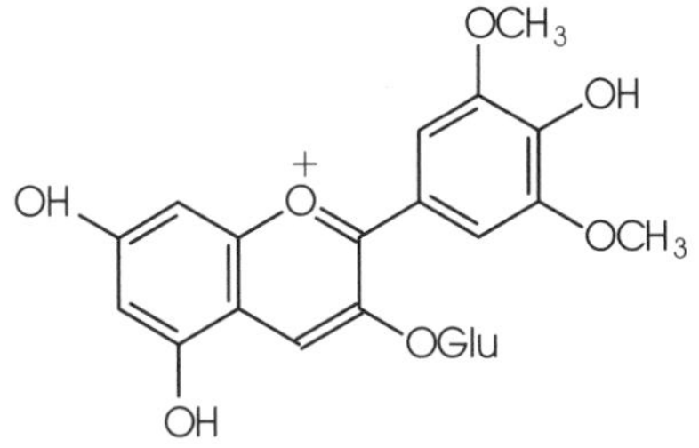

Figure 8.11 An example of a red flavylium cation, the colorant found in red wine. Glu represents the sugar glucose which has been linked to the main molecular skeleton. Other sugars produce other flavylium cations

for a particular class of organic cations.) The structure of one of this group of compounds is given in Figure 8.11. In red wine the flavylium cations are in equilibrium with other colourless and violet forms and only about 30 per cent of the anthocyanins present actually contribute to the initial red colour.

It is well known that the colour of red wine changes over time from an initial bright ruby red via purple red, plum and brick red to a pale tawny colour. While the chemistry of the changes is not fully understood, it is known that the overall cause is polymerisation of the flavylium cations. Within about 1 year of being made, about 50 per cent of all of the anthocyanin material is in the form of short polymer chains known as *oligomers*. Initially these molecules enhance the red colour of the wine because the conjugated bonding is more extensive in the oligomers than in the monomers. As the polymerisation increases the polymer tends to precipitate and the colour starts to fade, leading to the colour sequence mentioned.

8.9 Indicators

Indicators are molecules of weak organic acids which change colour as a function of the acidity (pH) of the surrounding aqueous solution. They are widely used in titrations to determine the progress of reaction between acidic and alkaline solutions.

The best known indicator, *litmus*, is a blue colouring matter derived from various lichens. It is chiefly composed of two compounds, *azolitmin* and *erythrolitmin* combined with alkalis. It becomes red in acid solution and blue in alkaline solution. Besides litmus there are a large number of other indicators, some of which are listed in Table 8.3, which operate over varying pH ranges and which display a variety of colour changes.

The reason for the colour change in an indicator is that some hydrogen atoms (*acidic* hydrogens) are lost or gained by the indicator molecule depending upon the pH of the solution. This hydrogen exchange causes a change in molecular structure, which, in the indicators, produces molecules of different colours. For example, some of the structures of colourless and red forms of the phenolphthalein molecule are given in Figure 8.12. The colourless form of this molecule is the acidic form, which includes a hydrogen atom in the structure. Removal of this hydrogen, which occurs in alkaline solutions, allows for greater resonance and a subsequent shift of the π to π^* absorption band into the blue region of the visible. The indicator then takes on a pink-red colour.

The point at which an indicator changes colour can be quantified in the following way. The general reactions taking place for an indicator in solution are:

$$HIn(aq) + H_2O(l) \rightarrow H_3O^+(aq) + In^-(aq)$$

where HIn represents the unionised form of the indicator and In^- the ionised form. The colour change is brought about because the ionised form is different in colour to the unionised form. The reaction can be treated by means of normal chemical equilibrium theory which allows us to write the expression:

$$K_a = [H_3O^+] \, [In^-] \, / \, [HIn]$$

where K_a is the acid dissociation constant of the indicator. The "end point" of a titration is arrived at when $[In^-]$ is equal to $[HIn]$. At this point the relationship

Table 8.3 The colours of some indicators

Indicator	Colour: acid	Colour: alkali	pK_a
Methyl orange	Red	Yellow	3.4
Bromophenol blue	Yellow	Blue	3.9
Bromocresol green	Yellow	Blue	4.7
Methyl red	Red	Yellow	5.0
Litmus[a]	Red	Blue	~7[a]
Bromothymol blue	Yellow	Blue	7.1
Thymol blue	Yellow	Blue	8.9
Phenolphthalein	Colourless	Pink	9.4
Alizarin	Red	Purple	11.7

[a] Litmus is a complex mixture of molecules, the principal indicator components of which are polymeric. For this reason litmus does not have a well defined value for K_a. It is useful for qualitative study, especially as litmus paper, but is not often used for quantitative work.

Figure 8.12 (a) The principal colourless (acid) form of the indicator molecule phenolphthalein. Loss of a proton from this structure, occurring in alkaline solution, leads to the formation of a number of resonance hybrids which are pink-red in colour, two of which are depicted in (b)

$$[H_3O^+] = K_a$$

holds. Now the pH of the solution is given by $\{-\log_{10}[H_3O^+]\}$ so that:

$$pH = pK_a$$

where pK_a is $\{-\log_{10}[K_a]\}$. Thus the colour of an indicator changes when the pH of the solution passes the pK_a value listed in the table. With many indicators the colour change is sharp enough for the end point to be gauged by eye to within one drop of added solution.

8.10 DYES

Most of the naturally occurring molecules discussed above can be considered to be dyes but most modern dyestuffs are synthetic chemicals. This is because commercial dyes must have certain properties apart from colour before they are useful. The dye must be *fixed* to the fabric in some way, and it must not be *fugitive*. These terms mean that the dye must be firmly attached to the fabrics and be stable with respect to light, and the normal conditions of washing. The actual mechanism by which a material becomes dyed is complex and depends upon both dye and fabric. All aspects of dyes and dying are the subject of extensive and continuing study.

There are over 7000 commercial dyes available, which go under more than five times as many trade names. Here we will only mention three of particular interest. *Mauveine* was the first commercially synthetic dye made, in 1856, and its production marked the birth of the synthetic dyestuffs industry. The discoverer, William Perkin, found that the material which he extracted via the oxidation of aniline sulphate could be used as a purple dye. Initially it was successfully used on silk under the name of Aniline Purple or Tyrian Purple[5]. In 1857 Perkin discovered how to apply this dye to cotton using tannin as a mordant, a compound used to attach the dye molecules to the fabric, leading to its very widespread use. In France the dye was extensively used under the name of *Mauve* and the

---

[5] This name was applied incorrectly, although it was used when the dye was first employed. Tyrian Purple (from Tyre, in Asia Minor), a natural dye obtained from certain shellfish, is 6,6-dibromo-indigo. It has a different structure to Mauveine. Tyrian Purple was both expensive and scarce, and in antiquity only the richest citizens, especially the Roman emperors, were in a position to wear garments stained with this exotic colorant.

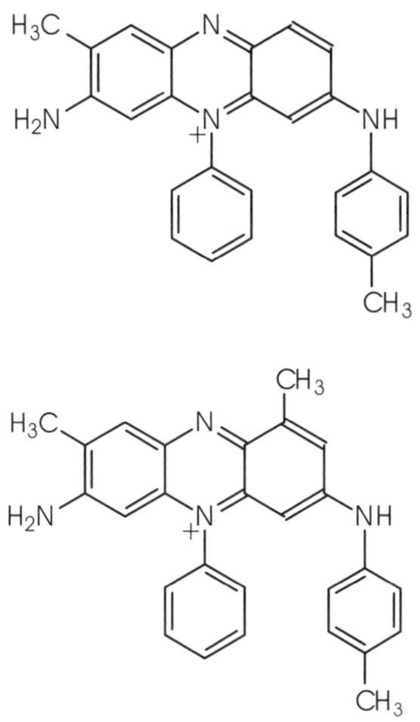

Figure 8.13 The dye mauveine, the first synthetic dyestuff prepared, is a complex mixture consisting of these two main components

commercial compound is now known as mauve or mauveine. In recent times it has been found that mauveine is actually a complex mixture of molecules. The structures of the two main components of the dye are shown in Figure 8.13.

Fluorescein, a fluorescent compound, is one of a remarkable family of coloured materials closely related to phenolphthalein. Fluorescene is yellow with an intense green fluorescence (see Chapter 9), which imparts to the molecule an unmistakable colour. It is widely used to colour safety garments and is the familiar yellow-green marker colour used to highlight passages of text. The structure, in Figure 8.14, can be compared to the similar phenolphthalein in Figure 8.12.

Indigo is one of the oldest dyes known to man. Indigo was a highly prized colouring material obtained many years ago from plants indigenous to Bengal, Java, and other parts of Asia. It is also the colouring in the plant woad. The invention of synthetic indigo had a severe economic effect upon the Asiatic indigo industry and led to considerable local hardship for many years. The structure of the dye indigo is drawn in Figure 8.15. The structure drawn, known as the *trans-* form, by virtue of the opposed nature of

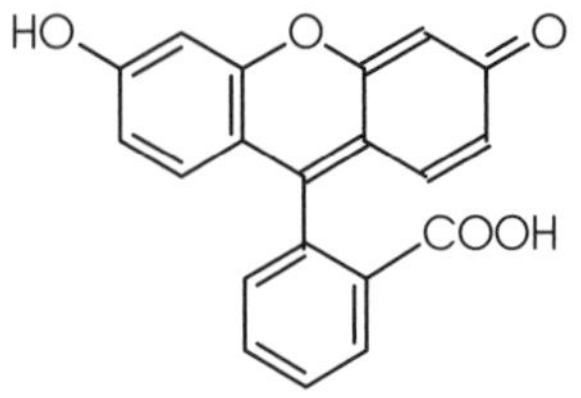

Figure 8.14 The structure of fluorescein, a yellow dye with a green fluorescence, widely used in safety garments

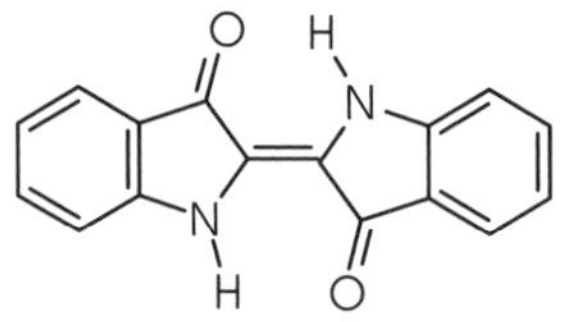

Figure 8.15 The *trans-* structure of the dye indigo, which occurs in both crystals and solutions, imparts the colour to "blue" jeans

the nitrogen atoms across the central double bond, is only one of several forms of the molecule, but is the one mainly responsible for colour production. Although the structure of indigo may not be well known, its colour, that of "blue" jeans, will be familiar to everyone.

8.11 ORGANIC PIGMENTS

Technically dyestuffs are soluble in the medium in which they are applied whereas pigments are insoluble. Pigments are thus used most frequently in a finely ground solid state, mixed with a carrier medium. They are used in this way in paints, inks and mixed with plastics to obtain opaque coloured products. Plate 8.3 shows a model of a crystal structure (of spinel, mentioned in Chapter 7 with respect to inorganic pigments), constructed from plastic polyhedra coloured by organic pigments.

In reality there is no fundamental difference between dyes and pigments. Many compounds can be used as dyes in one liquid and as a pigment in another. The phthalocyanines and the metal–organic complexes described earlier in this chapter are, for example, both important pigments and dyes.

8.12 COLOUR PHOTOGRAPHY

In both black and white and colour photography chemical processing changes small crystallites of silver halides which have been exposed to light into crystals of silver. The transformation hinges upon defects in the atomic structure of the halides which catalyse the reaction. In order to produce a coloured image the exposed silver grains, which yield a black and white image, must be replaced by suitable dyes.

One of the first widely used processes was that invented by Kodak and used to make Kodachrome transparencies (slides). The objective was to produce a transparent film containing an image built of dyes which absorb the complementary colours to those colouring the original. When the slide is projected using white light the complementary colours will be absorbed and colours of the original object will then be seen on a screen, as discussed in Chapter 1.

The colours of the original can be reconstructed by mixing three colours, red, green and blue. The three equivalent complementary colours needed in the finished slide are cyan, magenta and yellow, as in Table 1.2. These are produced in the emulsion during processing. The film emulsion is therefore produced in three layers.

(i) The top layer is sensitive to blue light and contains a dye precursor, referred to as a *dye coupler*, for the colour yellow.

(ii) This is followed by a layer of red sensitive emulsion containing the cyan dye coupler.

(iii) The final layer of emulsion is green sensitive and contains a magenta dye coupler.

The arrangement is illustrated in Figure 8.16a.

Let us consider, for example, a slide projected to show an area of blue sky. This part of the slide film must contain dyes able to remove the complementary colours to blue. This is achieved in the following way. When taking the photograph, blue light has reacted with the top layer of emulsion to form a latent image. The other two layers are designed so as not to react to the blue light and remain unchanged. When the film is developed the top layer of emulsion is converted into silver by chemical processing, as in Figure 8.16b. This silver is then removed chemically leaving the top layer clear and inert to light.

The next step during film processing is to *re-expose* the film to white light. (This step can now be carried out chemically in some processes.)

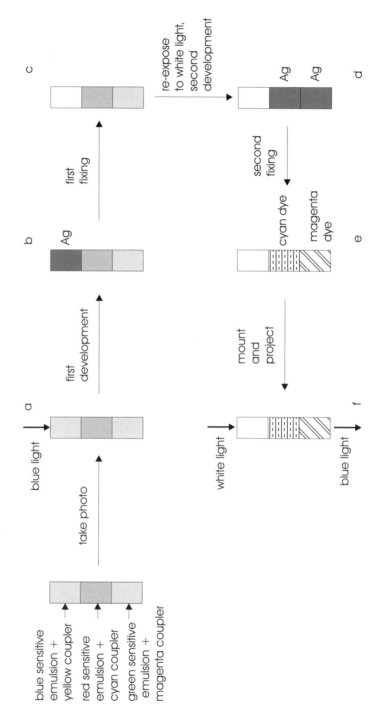

Figure 8.16 Processes taking place in the production of a blue colour slide. (a) Initial exposure of the three-layer emulsion to blue light. The processing steps (b–d) result in a transparency containing the complementary dye colours to that in the original light (e). Projection with white light (f) recreates a beam of blue light

This forms latent images in the other two layers. Development this time is designed to produce silver in the other two layers of emulsion, as in Figure 8.16c. At the same time silver ions in the remaining silver halides oxidise a portion of the developer.

$$\text{Developer} + 2Ag^+ \rightarrow \text{oxidised developer} + 2Ag + 2H^+$$

The oxidised form of the developer is reactive and combines with coupler molecules to form a dye. This takes place in both of the remaining two layers. The red sensitive layer will then contain a cyan dye and the green sensitive layer will contain a magenta dye. The remaining silver is removed chemically to leave a film containing cyan and magenta layers, as in Figure 8.16d.

After final removal of all unreacted chemicals, known as fixing, the process is complete. If white light is shone through the film, Figure 8.16f, the cyan dye will absorb the red component and the magenta dye will absorb the green component. Blue light will not be absorbed and will form an image on the screen.

The structure of the developer in its initial and oxidised form is given in Figure 8.17a. The dye couplers and dye molecules themselves are illustrated in Figure 8.17b, c, d. As in the other dye molecules described, conjugated bonds and resonance hybrid forms play an important part in producing deep colours.

8.13 Photochromic Compounds and Vision

A photochromic organic compound is a compound that undergoes a major reversible colour change, usually from colourless to deeply coloured, on irradiation with light. The reaction can be represented by the equation:

$$\text{A (colourless)} + h\nu_1 \rightarrow \text{B (coloured)}$$

As A is colourless it does not absorb in the visible and the ideal frequency for the activating photon is in the near ultraviolet. The reverse reaction takes place when the coloured form of the molecule absorbs light with a frequency near to the absorption maximum to yield the colourless product again.

$$\text{B (coloured)} + h\nu_2 \rightarrow \text{A (colourless)}$$

This second step is often known as "bleaching".

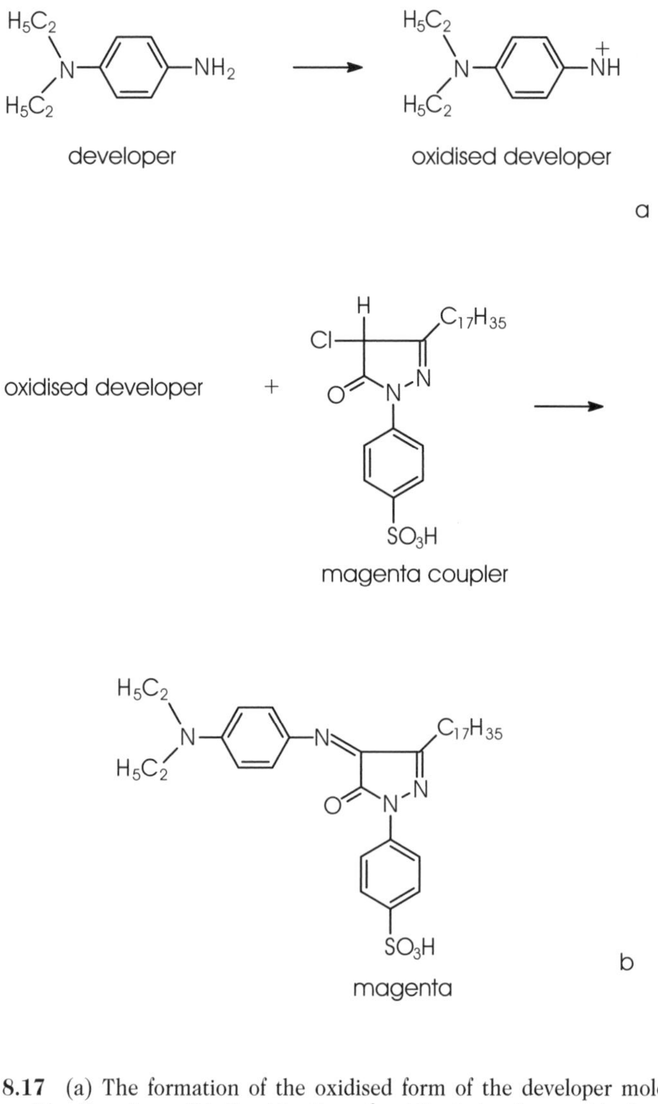

developer → oxidised developer

a

oxidised developer + magenta coupler →

magenta

b

Figure 8.17 (a) The formation of the oxidised form of the developer molecule by reaction with silver ions. (b) Combination of oxidised developer with dye coupler to produce magenta dye. (c) Combination of oxidised developer with dye coupler to produce cyan dye. (d) Combination of oxidised developer with dye coupler to produce yellow dye. The molecular structures are representative. Different manufacturers modify the colorants by changing some of the side groups attached to the molecular skeletons drawn

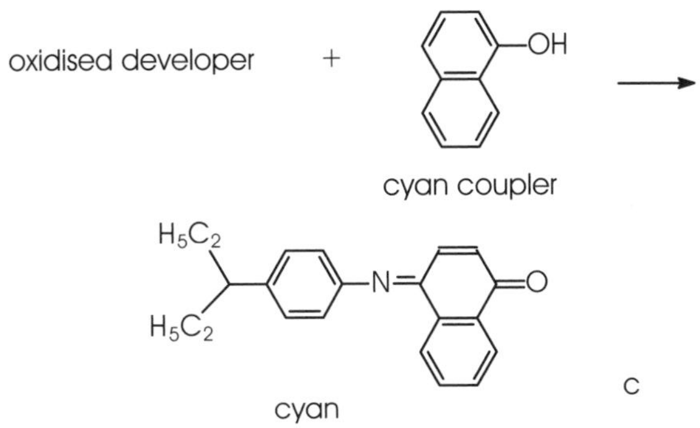

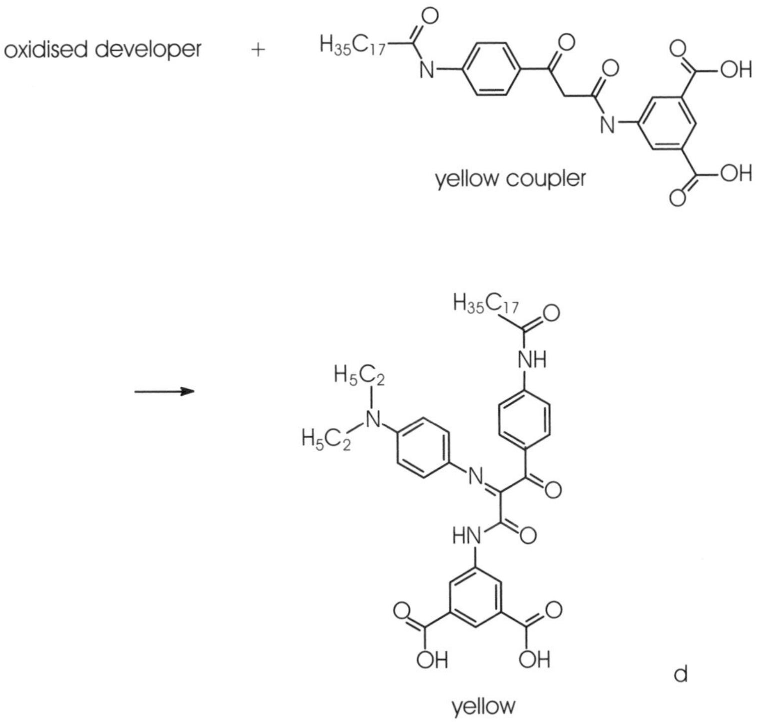

Figure 8.17 (continued)

The first photochromic reaction of an organic molecule to be reported, by ter Meer in 1876, was that of the potassium salt of dinitroethane, which changes from colourless to red in sunlight and back to colourless in the dark. Since the mid 1950s there has been a vast number of studies of photochromic molecules because of the potential for optical information storage. At present many hundreds of photochromic organic compounds are known. Despite the efforts of scientists engaged on these studies information storage systems using these materials have yet to be marketed, because of degradation reactions which damage the molecules during the repeated cycling from colourless to coloured forms. However, they have found uses in applications such as photochromic sunglasses.

Not long after ter Heer's observation, Boll, a German chemist, reported that the red-purple pigment found in the eyes of animals bleached in the presence of light to a colourless form. The change was found to be reversible, and in the dark the purple colour was regenerated. The compound became known as "visual purple" and is now called rhodopsin. This is undoubtedly the most important photochromic reaction for us as it is the source of our vision.

Vision in humans and other animals involves a complex set of photochromic reactions which take place in the rod and cone cells found in the retina of the eye. The rod cells are responsible for vision at low light intensities and are not sensitive to colour. The cone cells exist in three varieties, sensitive to either red, green or blue light and are responsible for colour vision. When light photons impinge on these cells they are absorbed by stacks of rhodopsin molecules which are bleached in the process. This sends a nerve impulse to the brain. The system is remarkably sensitive and there is considerable evidence to suggest that in the rod cells just one photon is enough to stimulate the nerve.

Not all of the structures and steps in the photochromic reactions taking place in the eye have been completely clarified as yet. Nevertheless, the framework of the processes is well established and is schematically outlined in Figure 8.18. Absorption of light by rhodopsin drives the molecule through several intermediates to the bleached state, which can consist of a number of different molecules, metarhodopsin I, metarhodopsin II and so on, depending upon the conditions experienced. These metarhodopsins can then revert, via other intermediate states to rhodopsin. Each cycle takes only a fraction of a second and can repeat indefinitely in normal light conditions so as to send a stream of nerve impulses to the brain. These end when the light is extinguished and all molecules revert to rhodopsin.

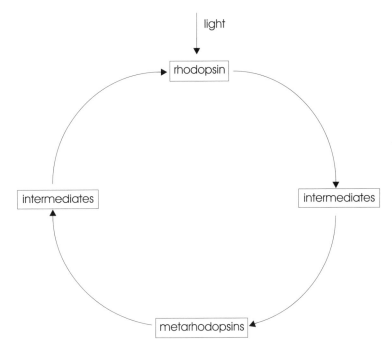

Figure 8.18 Schematic illustration of the cycle taking place when a rhodopsin molecule in the eye is activated by light. In normal illumination this process is repeated many times a second. Each cycle results in the transmission of a signal along the optic nerve to the brain

The chromophore of rhodopsin is the *cis*- form of the molecule *retinal*, sketched in Figure 8.19a, and properly labelled 11-*cis*-retinal. It is the aldehyde of vitamin A. This *cis*-retinal molecule is bound to the cell wall by a protein called an opsin via the amino acid lysine, to form rhodopsin. The *cis*-retinal by itself is not coloured and has an absorption maximum between 370 and 380 nm. However, when joined to the opsin the absorption maximum moves to about 570 nm. Molecules which can cause the deepening of the colour of a chromophore are called *bathochromes* and the resultant movement of the absorption maximum is referred to as a *bathochromic shift*. The bathochromic shift comes about because of the particular conformation of the *cis*-retinal molecule in conjunction with the protein. The *cis*-retinal fits into a pocket in the opsin surface which partly protects it from chemical attack by the environment. This is drawn schematically in Figure 8.20a.

We will first consider rod vision, which is not colour sensitive, under conditions of moderate incident light intensity. The molecular mechanism

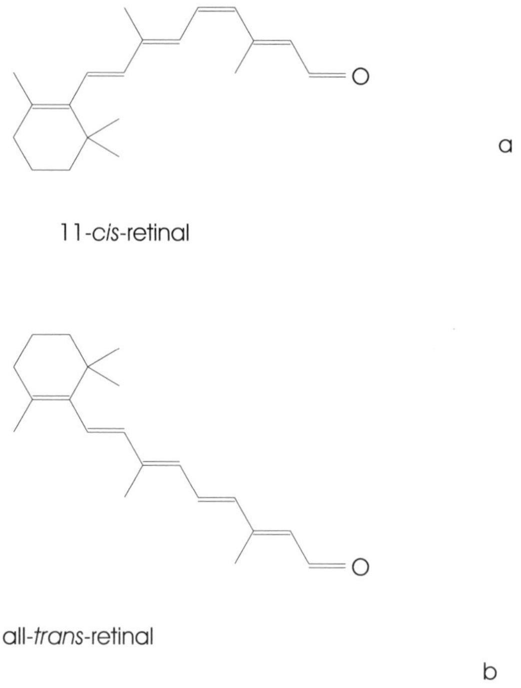

11-*cis*-retinal

a

all-*trans*-retinal

b

Figure 8.19 (a) The structure of 11-*cis*-retinal. (b) The structure of all-*trans*-retinal

leading to the nerve impulse hinges on the fact that retinal can exist in two isomeric forms, the *cis*- form already described, and a *trans*- form, called all-*trans*-retinal, drawn in Figure 8.19b. Under the influence of a photon the molecule changes from the *cis*- form to the *trans*- form. A number of complex steps then take place, via a number of intermediate stages, resulting in the formation of the molecule metarhodpsin I, as drawn diagrammatically in Figure 8.20b. This light induced change starts the chain of events which leads to the nerve impulse. Thereafter the reaction reverses, again passing through a number of intermediates, so that the *trans*-retinal readopts the *cis*- conformation and resumes its position in the fold on the opsin surface. Another photon can trigger the cycle again.

Although not yet completely clarified, it is thought that a similar mechanism to that described takes place in the cone cells. Here it is believed that the bonding of the retinal to different opsin molecules produces different bathochromic shifts and hence makes the cones sensitive to the different wavelengths of red, green and blue light.

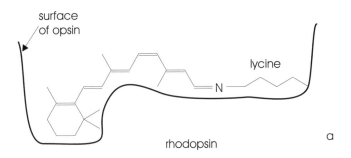

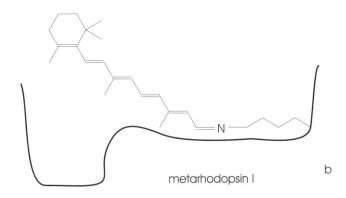

Figure 8.20 (a) A schematic illustration of the structure of rhodopsin, consisting of a *cis*-retinal molecule bound to the surface of an opsin by lycine. (b) The schematic structure of metarhodopsin I, in which *cis*-retinal has been transformed into all-*trans*-retinal by the action of a photon.

It is worth commenting on the enormous complexity of vision. The description of the cycle occurring in rod cells and presumed to occur in the cone cells described above is only true at moderate light intensities. At lower light intensities the *trans*-retinal molecule in rod cells is released completely from the opsin. Two processes then operate, dependent upon the weakness of the light signal. At the "higher" of these lower intensities the *trans*-retinal is transformed back to the *cis*-conformation by the action of enzymes in the eye itself, whereupon the molecule is re-attached to the opsin. At the lowest light intensities, the *trans*-molecules actually leave the eye completely, enter the bloodstream and are reprocessed to the *cis*- form in the liver, an occurrence which contributes to the length of time that it takes to become fully "dark adapted".

Rhodopsin has another role to play in the broader picture of life. It has been found that some purple halobacteria, bacteria which inhabit very salty environments, are coloured purple by rhodopsin. It is used in an analogous fashion to chlorophyll in plants. Absorption of light by chlorophyll initiates a chain of electron transfer reactions which eventually provides the energy for plant growth. In the purple halobacteria, the rhodopsin converts sunlight into energy for the metabolism of the bacterium. In essence it appears that the *cis-/trans-* change acts as a proton pump, and the resulting electro-chemical potential created initiates the energy building steps.

Much more about the interaction of light with organic molecules certainly remains to be discovered, both in the laboratory and in natural processes.

8.14 ANSWERS TO INTRODUCTORY QUESTIONS

Why is deep water tinted blue?

Water is blue due to molecular vibrations. These vibrations are mainly in the far infrared and in the case of isolated water molecules they do not encroach upon the visible spectrum. Thus water molecules in the atmosphere do not colour the sky in any way. In the case of liquid water or ice, the bonding between molecules is altered by an additional component: hydrogen bonding. This is sufficient to cause the highest energy portion of the complex absorption spectrum to extend slightly into the red end of the spectrum. This causes deep water (or blocks of ice), to take on a slight blue colour.

What colours roses red and cornflowers blue?

These colours are due to the presence of a group of compounds called anthocyanins. The colour of the molecules is influenced by the acidity of the sap in the plants which itself reflects the acidity of the soil. (The blue of hydrangeas is indicative of acid soil.)

Why does the colour of red wine change with age?

Red wine changes colour with age because the molecules which give the red colour slowly polymerise. The polymeric material precipitates out from the wine and collects in the bottom of the bottle. Thus red wines tend to lose colour and become more yellow (tawny) on ageing.

8.15 FURTHER READING

Much information on the structures and colours of organic molecules will be found in:

P. F. Gordon and P. Gregory, Organic Chemistry in Colour, Springer-Verlag, Berlin (1983).

J. Griffiths, Colour and Constitution of Organic Molecules, Academic Press, London (1976).

P. Rys and H. Zollinger, Fundamentals of the Chemistry and Applications of Dyes, Wiley, New York (1972).

The Merck Index, Merck, Rahway, USA. This handbook is updated and re-published at frequent intervals.

Details of analysis of metal ions using colour tests are given in:

F. Feigel and V. Anger, Spot Tests in Inorganic Analysis, Elsevier, Amsterdam (1971).

Earlier editions of this volume, author F. Feigl alone, are equally useful.

The photographic process is described, together with many references, in the reviews:

F. C. Brown, in Treatise on Solid State Chemistry, Vol. 4, Reactivity of Solids, ed. N. B. Hannay, Plenum, New York (1976).

J. F. Hamilton, Adv. Phys., **37**, 1988, p. 359.

Also, in technical literature produced by the manufacturers of film.

An excellent introduction to photochromic materials is given by:

H. G. Heller, Photochromics for the Future, in Electronic Materials, from Silicon to Organics, eds L. S. Miller and J. B. Mullin, Plenum, New York (1991), p. 471.

8.16 PROBLEMS AND EXERCISES

1. Would you expect "heavy water", D_2O, to be tinted like ordinary water?

2. Estimate the wavelength of the absorption maxima for the compounds $CH_3(CH=CH)_7CH_3$ and $CH_3(CH=CH)_9CH_3$ using the data for the absorption maxima for $CH_2=CH_2$ (λ = 162 nm), $CH_3(CH=CH)_4CH_3$ (λ = 296 nm), $CH_3(CH=CH)_8CH_3$ (λ = 405 nm). What colours would these two compounds be?

3. Why do the leaves of deciduous trees turn brown, red or yellow in autumn?

4. Copper containing compounds are often coloured blue or green (see Chapter 7). The blood of many crustaceans contains copper hemocyanin molecules. Is this why their blood is blue coloured?

5. Addition of the indicator methyl orange to a colourless solution turns it yellow. The addition of bromophenol blue has the same effect. What is the pH of the solution?

APPENDIX 8.1   RESONANCE HYBRIDS

In many molecules which contain conjugated double bonds the bonds between different atoms can be drawn in a number of alternative ways, each of which is equally likely to be correct from a chemical point of view. This concept is well represented by the molecule benzene, illustrated in Figure A8.1a,b. Originally the molecule was depicted as a hexagon of carbon atoms with three double bonds in the ring, which can be drawn in two ways. (In this representation the hydrogen atoms, one of which is joined to each carbon in the ring, are omitted.) Now in these circumstances neither bond arrangement is strictly correct. The real situation is a blend of the possible arrangements that can be drawn. The blended structure is called a *resonance hybrid*. The electrons which constitute the second bond in the double bond pair are spread over all of the molecule and are said to be delocalised and to occupy π *molecular orbitals*. Benzene is better drawn as in Figure A8.1c, where the hexagon shows the arrangement of carbon atoms linked by single bonds

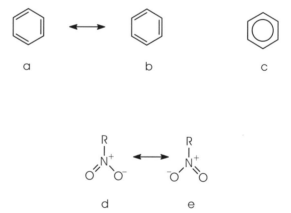

Figure A8.1   (a, b) Two resonance structures of benzene, C_6H_6. The hydrogen atoms have been omitted in this representation. The real structure of benzene is neither of these but a "blend" or resonance hybrid of the two, represented by (c). (d, e) Two resonance structures of the nitro- (-NO₂) group. R represents any organic or inorganic molecular fragment

and the delocalised electrons in a π molecular orbital are represented by the circle enclosed inside the hexagon. (The hydrogen atoms are again omitted.)

Resonance is not confined to organic molecules. Two resonance hybrids of the nitro- ($-NO_2$) group are also shown in Figure A8.1d, e. The real structure of the ($-NO_2$) group will be a blend of the two structures shown.

Molecules which have resonance hybrids are often found to be strongly coloured. Moreover, the more extensive the resonance (that is the larger the structure and the more resonance forms which can be drawn), the more intense the colour becomes. Almost all of the colorants and dyes discussed in this chapter fall into this pattern, although the different resonance structures have not been drawn.

It should be mentioned that the link between resonance structures and colour remains useful for the prediction of colour in organic molecules. The basic premise of this originally empirical idea has been convincingly supported by molecular orbital calculations which are now used to explore molecular colorants in a quantitative way.

CHAPTER 9

COLOUR FROM CHARGE TRANSFER AND LUMINESCENCE

Why do near colourless solutions of ferric chloride and potassium ferrocyanide become deep blue when mixed?
What produces the blue of sapphire?
How do fluorescent tubes produce white light?
By what basic process does up-conversion produce visible light?

In charge-transfer processes colour is produced by electronic transitions. A charge-transfer transition is one in which a relatively large redistribution of electron density occurs. This can occur in several ways. When two metal cations are involved the electron redistribution can involve electron transfer from one cation to another, in a cation-to-cation or intervalence transfer. Cations can also give or receive electrons from surrounding non-metal atoms in cation-to-ligand or ligand-to-cation processes. Finally, the electron redistibution might simply involve charge transfer between orbitals that are largely localised in different parts of the same molecule or complex. These transitions are similar to those that were discussed in the previous chapter and described as π-to-π^* and n-to-π^* transitions. Note, however, that not all of these latter transitions qualify as charge-transfer processes, only those in which the electron-donating group of atoms is quite distinct from the electron-accepting group and which result in a significantly redistributed charge density in the molecule. Charge-transfer transitions are depicted schematically in Figure 9.1.

Luminescence is somewhat similar to charge transfer because the energy resulting in colour is often absorbed by one centre and transferred to another. This second centre loses energy to produce colour.

In many of these colour producing mechanisms it is possible to write the transfer in ionic terms, and we will do this when it is reasonable. However, this simple approach does not always make good sense chemically as it sometimes entails the creation of unlikely ionic species. A better strategy is to construct a set of molecular orbital energy levels for the extended group

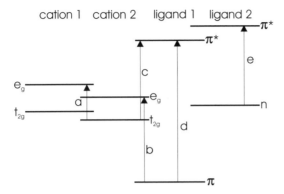

Figure 9.1 Some possible charge-transfer transitions in a metal complex. The metal d-orbitals have been assumed to adopt an octahedral crystal-field splitting. (a) Cation-to-cation. (b) Ligand-to-cation. (c) Cation-to-ligand. (d) π-to-π^*. (e) n-to-π

of atoms between which the charge transfer is occurring. The colour can then be considered to arise in electron transitions between the highest occupied molecular orbitals and the lowest unoccupied molecular orbitals, as in the case of the organic molecules discussed in the previous chapter.

9.1 CATION-TO-CATION OR INTERVALENCE CHARGE TRANSFER

Mixtures of metal ions can produce intense colours by the transfer of an electron from one cation to another. Such transitions are called *intervalence* or *cation–cation* charge-transfer transitions. For these to occur the cations must be able to adopt two different valence states. Moreover, if the ions are widely separated or if the anion geometry surrounding one cation is quite different from that surrounding the other the transition will not occur.

Charge-transfer colours are often observed in materials containing iron in both the Fe^{2+} and Fe^{3+} forms. In such transitions an electron is transferred from the Fe^{2+} to the Fe^{3+} ion.

$$Fe^{2+} [A] + Fe^{3+} [B] \rightarrow Fe^{3+} [A] + Fe^{2+} [B]$$

where [A] and [B] refer to the sites occupied by the ions. For such a transition to produce colour, some energy must be absorbed, which means that the final state must be at a different energy than the initial state.

One of the best known examples of Fe^{2+} – Fe^{3+} charge-transfer coloration is provided by the dark blue compound known as *Prussian blue* or *Turnbull's blue*. Prussian blue, long used as a pigment in inks, is a precipitate prepared by adding an aqueous solution of pale yellow $K_4[Fe^{2+}(CN)_6]$ to a pale yellow-green aqueous solution of any Fe^{3+} salt. Turnbull's blue, which seems to be the same chemically as Prussian blue, is made by mixing an equally pale coloured aqueous solution of $K_3[Fe^{3+}(CN)_6]$ with a pale green aqueous solution of an Fe^{2+} salt. The reaction in each case is quite spectacular. The mixing of two virtually colourless solutions instantly produces a dark blue-black colour.

The composition of Prussian blue is open to some uncertainty. One well investigated form, sometimes called "soluble" Prussian blue, has the formula $KFe_2(CN)_6$, and contains equal quantities of Fe^{2+} and Fe^{3+}. The basic structure of this compound is shown in Figure 9.2. It is seen that the Fe^{3+} and Fe^{2+} ions form a face-centred cubic array and the large K^+ cations occupy alternate cube centres. The linear CN^- ions, not shown in Figure 9.2, sit on lines midway between each of the Fe cations. Crystals also contain a variable amount of water in the structure. The energy of the structure depends sensitively upon the ordered arrangement of Fe^{2+}–Fe^{3+} ions. The process of changing an Fe^{2+} to an Fe^{3+} ion or *vice versa* will alter the lattice energy of the structure to some extent, which provides some of the energy difference between the initial and final states that was mentioned above.

A number of transition metal oxides, including titanium dioxide, TiO_2, become intense blue-black when very slightly reduced. The effect is quite noticeable, for example, in $TiO_{1.9993}$, in which only seven oxygen atoms in 20 000 have been lost. Although the exact details of the colour production has yet to be clarified, charge transfer between cations is the most likely candidate for coloration. This arises because loss of oxygen is balanced by a change in the valence of some of the cations present in the oxide. In TiO_2, for example, the parent oxide contains only Ti^{4+} ions. Each oxygen ion lost simultaneously causes two Ti^{3+} ions to form in the matrix. Cation-to-cation charge transfer can then be realised. The effect is used in the marking of cables in aircraft. Plate 9.1 shows the outer cover of wires marked by irradiation with intense ultraviolet light. In this case the plastic coatings contain significant amounts of titanium dioxide as an opacifier. The ultraviolet light, produced in a laser (see Chapter 13), is focused on the coating and is absorbed by the titanium dioxide. This causes a local loss of oxygen, resulting in an equally localised darkening of the material because of charge transfer.

Plate 8.1 A yellow water lily flower (*Nymphaea* hybrid Chromatella), coloured by flavonoid pigments, and a butterfly (*Maniola jurtina*), coloured by melanin-related pigments

Plate 8.2 (a) A rose flower

(b) A geranium flower. Both are coloured by anthocyanidin-related pigments. The colour difference is due to the different acidities of the sap in the plants

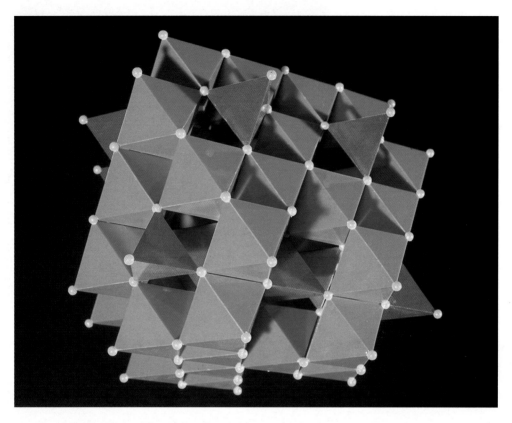

Plate 8.3 Plastic polyhedra brightly coloured by organic pigments. The model is of the crystal structure of the compound spinel, $MgAl_2O_4$

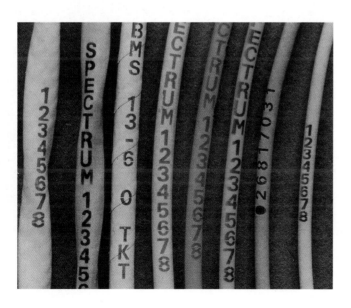

Plate 9.1 Plastic coatings on wires containing titanium dioxide, TiO_2, coloured by charge transfer, induced in the oxide by laser irradiation. Reproduced by permission of Spectrum Technologies plc

Plate 9.2 (a) Limestone rocks containing Fe^{3+} cations and coloured by charge tranfer between the cations and the surrounding oxygen ions

(b) A stream discoloured by deposits of iron oxy-hydroxides due to the transport of Fe^{2+} to the surface from disused mine workings. Reproduced by permission of Dr A. Eddington

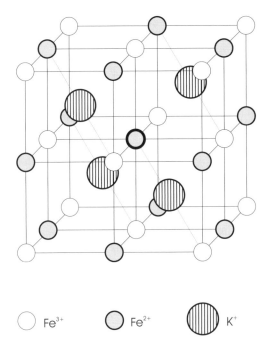

○ Fe^{3+} ◉ Fe^{2+} ◍ K^+

Figure 9.2   The crystal structure of Prussian blue, $KFe_2(CN)_6$. The $(CN)^-$ groups, not drawn, are found midway along the lines connecting the Fe ions so that each Fe is surrounded by an octahedron of $(CN)^-$ groups. The structure also contains a variable amount of water, which is not included

9.2   ELECTROCHROMIC FILMS

Tungsten trioxide, WO_3, is a pale lemon-yellow colour. Rather like titanium dioxide, it turns intensely blue-black when slightly reduced. The colour change is thought to be generated in a similar way to that described for titanium dioxide. That is, slight oxygen loss generates an additional valence state of tungsten in the solid, either W^{5+} or W^{4+}. Cation-to-cation charge transfer between a parent W^{6+} and a reduced ion can then colour the sample.

Although the colour change can be used to store information via laser marking, as in the case of titanium dioxide, there has been more interest given to using tungsten trioxide in dynamic displays. Unfortunately chemical reduction cannot easily be used for a display, but it has been found that the blue-black colour can also be induced when tungsten trioxide is sandwiched between electrodes carrying a voltage. Materials which change colour when subjected to an electric field are known as *electrochromic*. Tungsten trioxide falls into this category.

The mechanism of the electrochromic colour change involves a group of compounds called *tungsten bronzes*. These compounds are formed when hydrogen or alkali metal impurities enter the tungsten trioxide lattice. They are written M_xWO_3, where M represents the impurity atoms. Although the colour of the tungsten bronzes has not been explained fully, the colour induced at small values of x is so similar to that of reduced tungsten trioxide that it is presumed that charge transfer between two valence states of tungsten is occurring. This interpretation is supported by the fact that the colour is the same regardless of the impurity introduced. Moreover, the added impurities appear to be fully ionised and the electrons needed for charge balance are localised upon tungsten atoms at these low concentrations. (At higher concentrations the electrons become delocalised and the bronzes become metallic in behaviour and frequently brightly coloured. It is for this reason that they are known by the name of bronzes, although they are oxides, not alloys.)

Thin films of tungsten trioxide are transparent. At a crystallographic level, tungsten trioxide is constructed of corner-linked WO_6 octahedra. The result is a rather open structure and so it is quite easy to introduce a low concentration of hydrogen or metal atoms between the WO_6 octahedra to make blue-black tungsten bronzes. The principle of an electrochromic device using tungsten trioxide films is therefore not too difficult to envisage. Each pixel of the display would consist of a thin film of tungsten trioxide sandwiched between two electrodes, one of which should contain either alkali metal or hydrogen ions. An applied voltage can be set up so as to drive these ions into the tungsten trioxide, thus turning the film blue-black. Reversal of the voltage drives the impurity ions in the opposite direction back into the original electrode, causing the film to become colourless once more. This reverse process is often referred to as *bleaching*. The speed of coloration or bleaching depends mainly on ionic diffusion and at ordinary temperatures this is too slow for fast displays such as TV, but is satisfactory for electronic notice boards or shop signs. A similar arrangement has been used to make car mirrors which can be electrically dimmed so as to cut down dazzling reflections from bright lights. These cleverly use available water vapour to generate H^+ ions which in turn are used to produce a dark hydrogen tungsten bronze H_xWO_3 in the front surface of the mirror. This has the effect of reducing glare by reflection. Both displays and mirror systems are still in the development stage but show considerable potential for exploitation.

9.3 CONCENTRATION DEPENDENCE OF INTERVALENCE COLOURS

The colour of a charge-transfer material has been found to depend upon the concentration of ions present. When the concentration of the ions involved is low the charge-transfer bands give rise to less intense colours. For example, Fe^{2+}–Fe^{3+} charge-transfer transitions are responsible for the blue colour of *aquamarine*, which is a form of the mineral beryl containing small amounts of iron as an impurity. The blue colours become deeper and darker as the concentration of iron increases so that Prussian blue is, in fact, blue-black. The colour deepens even more in the black mineral *magnetite* or *lodestone*. This material has the spinel structure with a formula Fe_3O_4, or $(Fe^{3+})_t[Fe^{2+}\ Fe^{3+}]_oO_4$. Half of the Fe^{3+} cations in this structure are found in tetrahedral sites, written as $(Fe^{3+})_t$ and the remainder, together with the Fe^{2+} cations, are in octahedral sites, written as $[Fe^{2+}\ Fe^{3+}]_o$. Charge transfer does not occur between the ions on octahedral and tetrahedral sites because the change in geometry between the two sites is too large. However, it does occur between the ions which reside only on octahedral sites. Interactions between the iron ions at such high concentrations broadens the absorption band so much that all visible wavelengths are absorbed and the material looks black.

Intervalence transitions need not involve only one type of cation. The gemstone *sapphire* is formed from colourless, Al_2O_3, containing less than 1% of both Ti^{4+} and Fe^{2+}. These occupy neighbouring face-sharing octahedra in the structure. The charge transfer taking place is :

$$Fe^{2+}\ [A] + Ti^{4+}\ [B] \rightarrow Fe^{3+}\ [A] + Ti^{3+}\ [B]$$

As with the Fe^{2+}–Fe^{3+} transitions mentioned above, when the concentration of the cations becomes very high the beautiful blue colour is lost and the material becomes black. This occurs, for example, in the mineral *ilmenite*, $FeTiO_3$ which has a similar structure to Al_2O_3 but the Al^{3+} ions are replaced by an ordered arrangement of Fe and Ti. It is jet black in colour and occurs as black sands on beaches in several parts of the world.

9.4 ANION-TO-CATION CHARGE TRANSFER

Potassium permanganate, $KMnO_4$, forms dark purple, almost black, crystals. The crystals are only slightly soluble in water, but produce an intense purple coloured solution. The colour is associated with the $(MnO_4)^-$ ion, as K^+ ions never show colours in solution. Although it might be thought that the manganese alone could be responsible for the colour, this is not so. The manganese ion has a formal charge of Mn^{7+}, which indicates that it has lost all the d-electrons and so will not show crystal field colours. In addition, the absorption spectrum of the solution is quite unlike crystal field induced absorption. In fact the colour is attributed to a charge transfer between an oxygen ion in the $(MnO_4)^-$ group and the central Mn^{7+} ion. This is an *anion-to-cation* or *ligand-to-metal* charge transfer process.

A number of other transition-metal anions also show intense anion-to-cation charge-transfer colours. Among the most familiar is the dichromate ion, $(Cr_2O_7)^{2-}$, which gives crystals of potassium dichromate, $K_2Cr_2O_7$, a red colour in crystals and an intense orange-yellow colour in aqueous solutions. The bright colours of $PbCrO_4$, chrome yellow, and $BaCrO_4$, lemon yellow, also arise from similar ligand-to-metal charge transfer.

An anion-to-cation charge-transfer process is also responsible for the familiar yellow-brown colour of iron oxide, hematite, Fe_2O_3, and the various iron oxide-hydroxides that occur commonly in soils and rocks. The common red-brown colour of bricks, flower pots and many baked clay artefacts arises from the same source. As in the case of manganese, the crystal field colours of Fe^{3+} are too weak to account for the strong yellow-brown absorption of these iron compounds. The explanation is that we have charge transfer between O^{2-} or OH^- and Fe^{3+}. Plate 9.2a shows limestone containing Fe^{3+} ions. The familiar warm tones of the rock, caused by anion-to-cation charge transfer, is much prized in buildings.

It is this charge-transfer colour that is responsible for the discoloration of streams and rivers in old coal mining areas. Deep underground, fairly large amounts of FeS_2 exist within coal deposits. When mining operations cease water builds up in the workings and dissolves the sulphide to give Fe^{2+} ions in solution. These are eventually transported to the surface where they emerge as Fe^{2+} in streams. At this stage the water still looks clear. However, it rapidly becomes a bright yellow-brown colour because of the oxidation of Fe^{2+} to Fe^{3+} and the subsequent appearance of the charge-transfer colour of the hydrated Fe^{3+} species, as in Plate 9.2b. To make matters worse, the rather insoluble complex iron oxy-hydroxides formed are deposited as a glutinous mass on weeds and rocks. These not only look unattractive, but prevent the plants from continuing photosynthesis and clog the gills of many aquatic animals. In severe cases the result is a discoloured stream devoid of plant and animal life.

9.5 INTRA-ANION CHARGE TRANSFER

Although the blue colours derived from litmus, indigo and woad, mentioned in Chapter 8, were suitable for some coloration of fabrics, they were not found to be satisfactory for art work. This is because they are sensitive to pH changes and are also prone to lose colour. Paintings, from the middle ages until close to 1830, used very little blue at all, and the blues that were used tended to be produced from copper or cobalt compounds. These were also regarded as unsatisfactory by artists and only used reluctantly. There was, however, one exceptional blue pigment available, made from the mineral *lapis lazuli*. This is a rare blue stone found in isolated deposits mainly in Asia. Lengthy treatment of the mineral produced the fine blue pigment *ultramarine*. However, it was expensive (of the order of 10 000 French francs per kilo in 1830) and only manuscripts and paintings commissioned by the wealthiest of patrons, who also wished to advertise their wealth, used any large quantities of ultramarine.

The purple-blue colour in lapis lazuli is due to *lazurite*, an aluminosilicate with an approximate composition given by $(Ca,Na)_8(Al,Si)_{12}O_{24}(S, SO_4, Cl)_x$. The colour arises from the presence of a polysulphide anion with an approximate formula S_3^-. The unit consists of a triangle of three sulphur atoms together with one additional electron. The molecular orbitals of this cluster are not fully occupied and a transition between the filled and empty levels produces a strong absorption band at about 600 nm in the yellow region of the spectrum. (Note that the charge transfer occurs within this group of three sulphur atoms. It involves a redistribution of the charges within the S_3^- unit itself and not from one S_3^- group to another.) The colour reflected by ultramarine is thus blue with purple overtones. In natural lazurite and ultramarine the colour depends upon the exact amounts of calcium, sulphur, chlorine and sulphate present and in particular is deepened by increased calcium and sulphur content, which encourages S_3^- formation.

The cost of ultramarine was so high that the French Société d'Encouragement pour l'Industrie Nationale and the British Royal Society of Arts both set up prizes for the discovery of an artificial method of ultramarine fabrication. A process using the easily obtainable clay kaolin was discovered in 1828 and from that time ultramarine has not been excessively expensive. The approximate equation of formation is:

$$6Al_2Si_2O_7.8H_2O \text{ (kaolin)} + 5Na_2CO_3 + 2Na_2SO_4 + 2CaCO_3 + 6S + C \rightarrow$$
$$2Na_7CaAl_6Si_6S_3(SO_4)O_{24} \text{ (ultramarine)} + 48H_2O + 7CO_2 + CO$$

If there is insufficient sulphur present a green colour, arising from S_2^- will appear. The production of a relatively cheap blue colouring was an important factor in the blossoming of the Impressionist movement of painters, and many of the classic paintings of this style contain copious quantities of synthetic ultramarine.

Many other polysulphides are coloured. The formation of extended groups of sulphur atoms gives rise to molecular orbitals which can participate in the charge transfer redistribution of electrons, and in so doing generate intense absorption colours.

9.6 LUMINESCENCE AND PHOSPHORS

Luminescence is a general term for the emission of radiation in the visible part of the spectrum by relatively cool bodies, as opposed to incandescence, which applies to light emission from hot bodies. *Phosphors* are materials which convert radiation of one wavelength to radiation of another wavelength, the conversion being from a higher energy to a lower energy. They are widely used in, for example, fluorescent lamps (ultraviolet to visible), TV (electron impact to visible) and scintillators (X-rays and γ-rays to visible).

Fluorescence refers to the *instantaneous* emission of radiation after excitation by some other external radiation. The radiation emitted has a longer wavelength (lower energy) than the exciting radiation. The "missing energy" is lost via *non-radiative transitions* (also called *phonon-assisted transitions*), which increase lattice vibrations, resulting in a warming of the fluorescent material. *Phosphorescence* is the *delayed* emission of radiation, but is otherwise identical to fluorescence. The delay can vary from milliseconds to hours or days. In this phenomenon, the absorbed energy can be thought of as stored in a reservoir from which it slowly leaks; a feature more commonly associated with heavy atoms. Indeed, Becquerel was studying the phosphorescence of uranium compounds when he discovered radioactivity.

The processes taking place in a phosphor can be illustrated with reference to Figure 9.3. The (usually inactive) matrix has incorporated within it an *activator*, A, which is able to absorb an excitation in the form of a photon of energy $h\nu_1$, and which can re-emit it as a photon of energy $h\nu_2$ or lose energy via internal processes. Sometimes A cannot absorb the photons directly, in which case a *sensitisor*, S, is added as well. In this case the sensitisor absorbs the exciting photons, of energy $h\nu_3$ and passes the

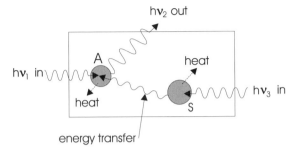

Figure 9.3 The energy absorption and emission processes taking place in a fluorescent material. A represents an activator centre and S a sensitisor centre. The photons absorbed and emitted are $h\nu_1$, $h\nu_2$ and $h\nu_3$. Some energy is also usually lost to the matrix as heat

energy to A, re-emits radiation itself or loses energy via internal transitions. S can be the crystal matrix itself and $h\nu_1$ can be identical to $h\nu_2$ and $h\nu_3$, although the emitted radiation, $h\nu_2$ is usually at a lower energy than that absorbed, $h\nu_1$ or $h\nu_3$, as some energy is generally lost to the matrix phase in the form of heat.

9.7 PHOTOLUMINESCENCE AND FLUORESCENT LAMPS

Photoluminescence is the process in which ultraviolet radiation fairly close in wavelength to the visible is converted into visible radiation. An example of green-yellow fluorescence of a zinc sulphide-based phosphor when irradiated with ultraviolet light is shown in Plate 9.3. The commonest example is fluorescent lighting. In these lamps, mercury vapour at a low pressure is used. Under electron bombardment the Hg atoms are excited and emit copious ultraviolet radiation with a wavelength of 254 nm as well as some radiation in the visible, as was explained in Chapter 7. Conversion of the ultraviolet radiation to visible is by way of a phosphor coated onto the inside of the tube, as depicted in Figure 9.4. The most widely used host matrix is calcium fluorophosphate, $Ca_5F(PO_4)_3$. When doped with Sb^{3+} ions as activator a light blue emission is produced. The Sb^{3+} ions have a good absorption band, due to an s^2 to s^1p^1 transition, at 254nm, which closely matches the mercury vapour output. A minor problem with Sb^{3+} is that the blue emission gives the lamps a rather cool colour. If Mn^{2+} is also incorporated into the system as a *coactivator* a warmer tone is produced as this ion produces an orange-red emission. Variation in the proportions of Sb to Mn varies the tone of the light. A typical emission spectrum from a

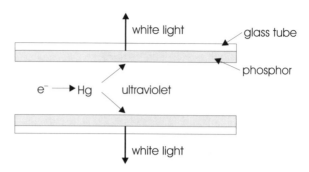

Figure 9.4. The processes occurring in a fluorescent tube. Electrons, e⁻, from the cathode collide with mercury (Hg) atoms which in turn emit ultraviolet radiation. This is converted into visible light by a phosphor coating on the inside of the glass tube

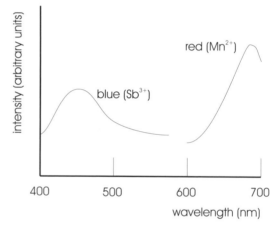

Figure 9.5 Schematic representation of the emission spectra from Sb^{3+} (blue emission) and Mn^{2+} (red emission) in a typical fluorescent tube phosphor

fluorescent tube phosphor is shown in Figure 9.5. Note that the emission bands are very broad because the orbitals involved in the electron transitions producing the light output can interact strongly with the surrounding crystal matrix.

There are a number of other aspects of the phosphor which are of interest. First, although the Mn^{2+} has good emission characteristics, it is found to be unsatisfactory when used alone. Fortunately the Sb^{3+} acts as a sensitisor for the Mn^{2+} ion, thus avoiding another component in the phosphor. Second, the Mn^{2+} and Sb^{3+} ions occupy the Ca^{2+} positions in the host matrix. Now while Mn^{2+} incorporation will not pose a problem, as the Mn^{2+} ions have the same charge as the Ca^{2+} ions that they replace, this is not so with Sb^{3+}. The introduction of Sb^{3+} ions into the phosphate will thus cause an internal charge imbalance which will result in a degradation of performance. To overcome this, charge balance is maintained by adding one F^- or Cl^- ion to the phosphate for each Sb^{3+} ion. It has been found that an empirically derived composition for the host matrix of $Ca_{10}P_6F_{1.8}Cl_{0.2}O_{24}$ is most satisfactory.

Over the years improvements have occurred in fluorescent lighting, especially with respect to the colour of the light produced. *Trichromatic* lamps produce a very good spectral balance by using a phosphor mixture which emits equal amounts of the colours red, blue and green. The favoured red emitter is Eu^{3+} doped into a Y_2O_3 matrix. A charge transfer band between Eu^{3+} and O^{2-} absorbs efficiently at 254 nm and accordingly this ionic pair can readily take up the ultraviolet radiation given off by the excited mercury atoms. Energy is then lost to the surrounding matrix as non-radiative transitions until the Eu^{3+} ion reaches the upper f-electron energy levels. At this point energy is lost as photons at a wavelength of 612 nm, allowing the return of the ion to the ground state. The green emission is from Tb^{3+}. This needs a sensitisor, usually Ce^{3+}, which is able to absorb the 254 nm wavelength mercury radiation. Typical host matrices are $La(Ce)PO_4$, $La(Ce)MgAl_{11}O_{19}$, and $La(Ce)MgB_5O_{10}$. In each case the Tb^{3+} and Ce^{3+} ions replace La^{3+} ions and no charge compensation is needed. The blue emission is produced by Eu^{2+} ions. The ultraviolet radiation is absorbed efficiently by this ion, utilising a transition from a 4f energy level to a 5d energy level. Decay by phonon and then photon emission, as with Eu^{3+}, yields the desired blue output. Among the most widely used host lattices for Eu^{2+} ions are $BaMgAl_{11}O_{19}$ and $Sr_4Al_{14}O_{25}$. An emission spectrum from a trichromatic fluorescent tube is illustrated in Figure 9.6. The emission lines are narrow compared to those in Figure 9.5 because the f-orbitals involved in the process are shielded from the surrounding matrix, as explained in Chapter 7.

9.8 UP-CONVERSION

The detection and display of infrared radiation has many possible applications, ranging from the study of nocturnal mammals to night combat. An interesting method of achieving this objective is variously known as *frequency up-conversion, anti-Stokes fluorescence* or *cooperative luminescence*. In this effect low energy radiation, typically in the infrared, is "up-converted" into visible radiation. The majority of studies have dealt

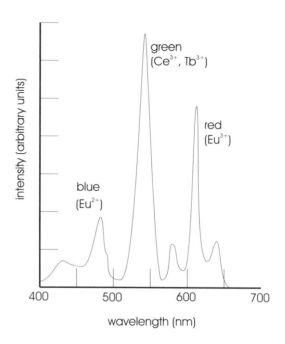

Figure 9.6 Schematic representation of the emission spectra from Eu^{2+} (blue emission), Tb^{3+}, Ce^{3+} (green emission) and Eu^{3+} (red emission) in a trichromatic fluorescent lamp

with the behaviour of polycrystalline or glass phosphors containing a rare earth couple, such as Yb^{3+}/Er^{3+} and we will describe the process with respect to this ionic pair. The phenomenon is distinct from that in frequency doubling. In up-conversion impurity lanthanide ions absorb and emit the radiation. In frequency doubling, undoped crystals are used and the frequency change is due to electronic polarisation, as described in Section 3.10.

The conversion takes place in two distinct *steps*. A long wavelength photon in the infrared, with a wavelength of about 1000 nm, is absorbed by a Yb^{3+} ion. The energy is then immediately transferred to a nearby Er^{3+} ion, as illustrated in Figure 9.7a. Before the excited Er^{3+} ion has time to lose energy the same thing happens again, and a Yb^{3+} ion makes another energy transfer to the same, now excited, Er^{3+} ion. This will doubly excite the Er^{3+} ion to twice the energy of the input photons, as in Figure 9.7b. This process is known as *optical pumping*. The doubly excited ion then falls back to the ground state and releases a photon in the green region, with a wavelength of 550 nm, shown in Figure 9.7c. An emission spectrum of the

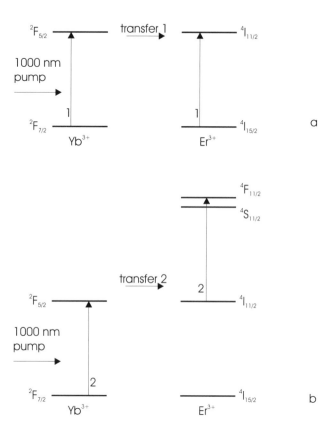

Figure 9.7 The steps involved in the up-conversion of infrared light with a wavelength of approximately 1000 nm to green light with a wavelength of 550 nm using a Yb^{3+}/Er^{3+} ion pair. (a) Absorption of the first infrared photon by Yb^{3+} and transfer to an Er^{3+} ion. (b) Absorption of second infrared photon by Yb^{3+} and transfer to the same (now excited) Er^{3+} ion. (c) Emission of green light by the doubly excited Er^{3+} ion as it returns to the ground state

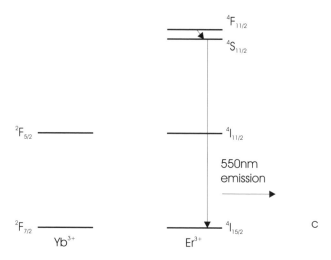

Figure 9.7 (*continued*)

output from a glass containing Yb^{3+}/Er^{3+} is shown in Figure 9.8. This also indicates a small output of red light which arises from an intermediate energy level not shown in the figure.

The *up-conversion efficiency* of this system can be defined as the ratio:

Efficiency = power emitted (green) / power absorbed (infrared)

In general the up-conversion efficiency is low and varies with the concentration of the activator and sensitisor ions. A maximum efficiency is observed with concentrations of about 5 per cent. Above this value the efficiency is limited by increasing back transfer from Er^{3+} to Yb^{3+} and increasing Yb–Yb energy transfer.

Two other up-conversion processes are known: (i) blue emission from a Yb^{3+}/Tm^{3+} couple using irradiation with a 1 μm wavelength infrared beam. The 1 μm pump wavelength is transformed by way of a triple excitation up-conversion process; (ii) red emission using Er^{3+} alone and a 1.5 μm pump wavelength. Note that this red generation is also to be seen in Figure 9.8 but the relevant energy levels are not included in Figure 9.7.

Two-frequency up-conversion expands upon the process just described and allows for the formation of three-dimensional images in a transparent glass matrix. Once again, lanthanide ions are involved, on this occasion praseodymium ions, Pr^{3+}, in a fluoride glass matrix. A simplified energy

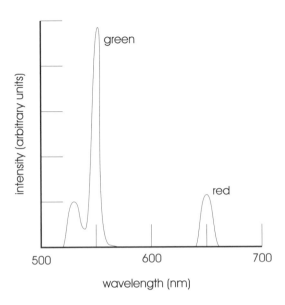

Figure 9.8 The emission spectrum of a glass containing a low concentration of Yb^{3+}/ Er^{3+} ion pairs irradiated with infrared light with a wavelength of approximately 1000 nm. The principal up-conversion is to the green, but a significant amount of red light from the Er^{3+} ions alone is also generated

level diagram of the ion is given in Figure 9.9. If the matrix is illuminated with infrared radiation of 1014 nm wavelength every active Pr^{3+} ion is excited to the metastable $^1G_4$ energy level. No further excitation is possible using the 1014 nm pump. However, if the excited ions are now irradiated with infrared radiation of 850 nm wavelength they can pick up energy and move to the $^3P_2$ state. These doubly excited ions rapidly decay non-radiatively to the $^3P_0$ state, then emit red light before finally returning to the ground state.

In this double excitation process the beam of radiation at 850 nm only interacts with already excited Pr^{3+} ions and no others. Similarly, irradiation of the material with a beam of infrared radiation at 1014 nm does not result in a visible emission. Up-conversion and visible output only takes place at the intersection of the two beams. This allows for the formation of a three-dimensional coloured image within the glass matrix.

Up-conversion processes have an advantage over frequency doubling using nonlinear crystals, described in Chapter 3, in that glasses can be used as the host matrix and expensive precisely oriented single crystals are not required. Although commercially available systems using up-conversion

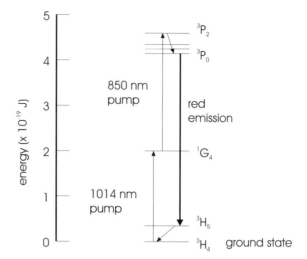

Figure 9.9 Simplified energy level diagram for Pr^{3+} ions in a fluoride glass. Two-frequency up-conversion excites these ions in two steps from the ground state to the $^3P_2$ level. Non-radiative decay lowers the energy to the $^3P_0$ level. Red light is emitted as the ions drop from the $^3P_0$ energy level to the $^3H_5$ energy level

are not yet available, there is sure to be considerable progress in this interesting area of research over the coming years.

9.9 Answers to Introductory Questions

Why do almost colourless solutions of ferric chloride and potassium ferrocyanide become deep blue when mixed?
The reason is that Prussian blue forms. In this compound Fe^{2+} and Fe^{3+} ions coexist. The blue colour is a charge-transfer colour between these two ions.

What produces the blue of sapphire?
Sapphire is Al_2O_3 (corundum) containing about 1% Ti^{4+} and Fe^{2+} on adjacent octahedral sites. The colour is a charge-transfer colour between these two ions.

How do fluorescent tubes produce white light?
These devices consist of a coating of phosphor on the inside of a glass tube. Mercury vapour within the tube emits copious amounts of ultraviolet

radiation. This is absorbed by Sb^{3+} and Mn^{2+} centres in the phosphor matrix and re-radiated at lower energy, in the visible.

By what basic process does up-conversion produce visible light?
The basic process which occurs is the adding up of low energy photons by a lanthanide ion in a suitable matrix. The addition can be thought of as each incoming low energy photon exciting the lanthanide ion up successive steps on an energy level ladder. In up-conversion the steps on the ladder are equally spaced and only bombardment by photons of one wavelength is needed. Visible light is emitted when the doubly or triply excited lanthanide ion returns to the ground state in one step.

9.10 FURTHER READING

Charge transfer colours are discussed in:

K. Nassau, The Physics and Chemistry of Colour, Wiley-Interscience, New York (1983), Ch. 7.

Information on luminescence and phosphors is in:

C. G. A. Hill, Chemistry in Britain, **19**, 1983, p. 723.
G. Walker, Chemistry in Britain, **19**, 1983, p. 80.
G. V. Subba Rao, in Perspectives in Solid State Chemistry, ed. K. J. Rao, Narosa, New Delhi (1995), p. 366.
G. Blasse and B. C. Grabmaier, Luminescent Materials, Springer-Verlag, Berlin (1994).

More on up-conversion can be found in the following two papers and the references therein:

J. Wojciechowski, I. Pawelska, R. Grodecki and L. Szymanski, J. Electrochem. Soc., **122**, 1975, p. 312.
W. Xu, G. Chen and J. R. Peterson, J. Solid State Chem., **115**, 1995, p. 71.
M. J. Dejneka, MRS Bulletin, **23**, November 1998, p. 57.

The defect chemistry which is involved in doping is explained in:

R. J. D. Tilley, Principles and Applications of Chemical Defects, Stanley Thornes, Cheltenham (1998), Ch. 7.

9.11 PROBLEMS AND EXERCISES

1. The three halides of silver show a gradual deepening of colour thus:
    AgCl, white, AgBr, pale yellow, AgI, yellow. Why might these colours
    occur?

2. $AuCl_3$ is red and AuCl is yellow. A simple halide $AuCl_2$ does not exist,
    but there is a compound $Cs_2Au_2Cl_6$ whose formula seems to suggest the
    presence of Au^{2+}. However, it is black. What conclusions can you draw
    about the state of the Au ions and the cause of the black colour?

3. Most high-T_c oxide superconductors containing Cu ions, including the
    well known material $YBa_2Cu_3O_7$, are black. Why might this be so?

4. If lemon-yellow WO_3 is very slightly reduced it becomes a deep blue-
    black. How might this change come about?

5. Three thallium tellurium oxides show the following colours: Tl_2TeO_3,
    lemon yellow, Tl_6TeO_6, orange, Tl_2TeO_6, black. What might be the
    cause of the black colour?

6. How much energy has to be degraded as heat before an Eu^{3+} ion in a
    phosphor mixture which has absorbed light at 254 nm can emit red
    light at 612 nm to return to the ground state?

CHAPTER 10
COLOUR IN METALS, SEMICONDUCTORS AND INSULATORS

Why are metals shiny in bulk, but black as powders?
Why are gold and copper coloured?
What produces the intense colours in the paints
vermilion and cadmium yellow?
How can colourless impurities turn diamonds blue
or yellow?
In what way do holes change quartz into amethyst?

10.1 BAND THEORY

In order to answer the questions posed above it is necessary to know something of the way electrons are held in metals, semiconductors and insulators. One of the best theories for this purpose is *band theory*. In this approach the electron energy levels are shown to be broadened into *energy bands*. The way this comes about is easy to understand. In an isolated atom the electrons occupy a ladder of sharp energy levels. In Figure 10.1a the outermost energy level of an isolated atom is shown. If another atom approaches the first the outer electron clouds will interact and the result is that the single energy level will split up into two, one at a higher energy and one at a lower energy, as shown in Figure 10.1b. Four atoms will give four energy levels, as illustrated in Figure 10.1c. This same process, in fact, gives rise to the molecular orbitals initially introduced in Chapter 8.

This process can be continued indefinitely. As each atom is added to the cluster the number of energy levels in the high energy and low energy groups increases. At the same time the spacing between the energy levels in each group decreases. Ultimately, when a large number of atoms are brought together, as in a solid, the energy levels in both the high energy and low energy groups are very close indeed. They are now called *energy bands* and are shown shaded in Figure 10.1d.

The details of the band structure of a crystal depend upon both the geometry of the structure and the degree of interaction of the electron

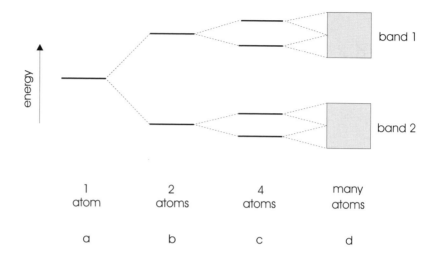

Figure 10.1 The formation of energy bands. (a) Isolated atoms have sharp energy levels. (b) Each energy level in a single atom becomes two energy levels (molecular orbitals) in a diatomic molecule. (c, d) As the number of atoms increases the number of energy levels increases until bands of very closely spaced energy levels form

energy levels. If the interaction is large, typically for the outer orbitals of closely spaced large atoms, the bands are broad. When the interaction is less, as occurs for inner electron orbitals on atoms which are further apart, the width of each band is rather narrow.

The electrons in the solid fill the bands from the lowest energy to the highest. The topmost band can be partly or completely full. Colours in these materials arise by electron transitions within and between bands, as we shall see.

10.2 INSULATORS, SEMICONDUCTORS AND METALS

Insulators have the upper energy band completely empty and the lower energy band completely filled by electrons as illustrated in Figure 10.2a. Moreover, the energy gap between the top of the filled band and the bottom of the empty band is quite large. The *filled* energy band is called the *valence band* and the empty energy band is called the *conduction band*. The energy difference between the top of the valence band and the bottom of the conduction band is called the *band gap*.

Intrinsic semiconductors have a similar band picture to insulators except that the separation of the empty and filled energy bands is small, as

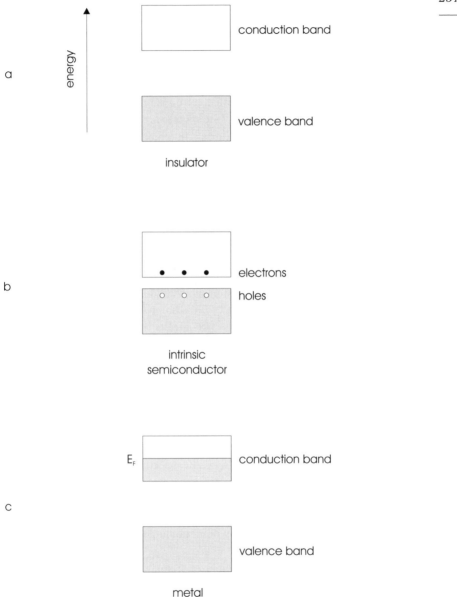

Figure 10.2 Schematic illustration of the energy bands in (a) an insulator, (b) an intrinsic semiconductor and (c) a metal. E_F represents the Fermi energy. The bands are idealised and do not show the three-dimensional band geometry which varies with direction in real crystals

in Figure 10.2b. How small is small? The band gap must be such that some electrons have enough energy to be transferred from the top of the valence band to the bottom of the conduction band at room temperature. Each electron transferred will leave behind a "vacancy" in the valence band. Rather surprisingly, these vacancies behave as if they were positively charged electrons. They are known as *positive holes*, or more often just as *holes*. Therefore, each time an electron is removed from the valence band to the conduction band *two* mobile species are created, an *electron* and a *hole*.

Metals are defined as materials in which the uppermost energy band is only partly filled as shown in Figure 10.2c. The highest energy attained by electrons in this band is called the *Fermi energy* or *Fermi level*.

10.3 THE REFLECTIVITY OF METALS AND THE COLOUR OF COPPER AND GOLD

The higher electronic energy levels of a metal are characterised by an essentially continuous band of allowed energies. Above the Fermi energy almost all the levels are empty (at absolute zero they are all empty) and so can accept excited electrons. This means that all incident radiation can be absorbed, irrespective of its wavelength, because there is always an empty level available to accept the excited electron, as shown in Figure 10.3.

Intuitively this would lead one to expect that a metal should appear black. However, each excited electron can immediately fall back to the state that it came from at once, emitting exactly the same energy, causing a flat piece of metal to appear reflective. Ordinary mirrors are metal films deposited onto glass. In a good mirror the absorption and reflection should be identical over the whole of the spectrum and all colours accurately reflected. Interestingly, exactly the same absorption and emission processes lead to finely powdered metals having a black appearance. This is because the re-emitted (i.e. "reflected") photons are re-absorbed again in nearby grains and ultimately do not emerge at the "angle of reflection" and so do not enter the eye.

In the case of insulators electronic polarisability is the main contributor to the refractive index in the visible range. However this is not so when the material becomes strongly absorbing, as was mentioned in Section 4.9. Absorption is also of importance in metals, because the electrons present in the conduction band also strongly absorb light. To take this into account, the refractive index N, of an absorbing material such as a metal, must always be written as:

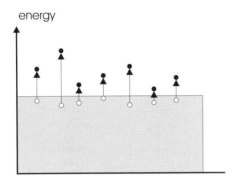

Figure 10.3 All photons incident upon a metal can be absorbed because there is an effective continuum of energy levels available above the top of the filled levels (shaded). The same photons are re-emitted when the excited electrons drop back to the lower energy levels

$$N = n + ik$$

where n is the "normal" refractive index defined in Chapter 2 and k is termed the *extinction coefficient, coefficient of absorption* or *attenuation coefficient*. The symbol i represents $\sqrt{-1}$. The reflectivity of a metal depends upon n, k and the polarisation of the light. For light falling perpendicularly on a metal surface, polarisation can be ignored and the reflectivity is given by:

$$R = [(n - 1)^2 + k^2] / [(n + 1)^2 + k^2]$$

If k is omitted the formula reduces to that for a normal insulator such as glass. For a metal the extinction coefficient k and the refractive index, n, are both strongly wavelength dependent. Some values are given in Table 10.1.

The colours of copper and gold are due to the fact that the absorption and emission of photons are dependent on wavelength. Table 10.1 and Figure 10.4 indicate that both gold and copper have rather low reflectivity at the short wavelength end of the spectrum and so yellow and red will consequently be reflected to a greater degree. This leads to the colours observed. Aluminium, on the other hand has a high reflectivity which does not vary significantly with wavelength, making it suitable for use in mirrors.

Table 10.1 Reflectivity of gold, copper and aluminium

Metal	λ (nm)	Refractive index (n)	Extinction coefficient (k)	Reflectivity (R)
Gold	689	0.09	3.82	0.9777
	564	0.24	2.54	0.880
	459	1.43	1.72	0.355
Copper	689	0.21	4.04	0.952
	564	0.83	2.60	0.673
	459	1.16	2.43	0.561
Aluminium	689	1.741	8.205	0.907
	564	1.018	6.846	0.920
	459	0.647	5.593	0.924

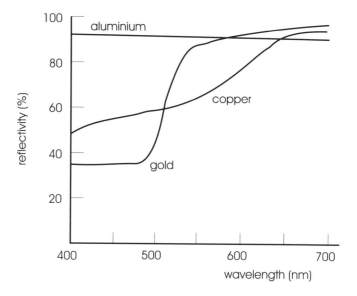

Figure 10.4 The reflectivity of aluminium, copper, and gold, as a function of wavelength, for light at normal incidence to the surface

Thin flakes of a ductile metal such as aluminium, produced by ball milling, are added to paints to obtain a "metallic" effect. Aluminium is especially suitable from this point of view and is the commonest metal used but bronzes and copper alloys are also employed for this purpose. They are rarely used alone but usually in conjunction with other pigments to produce a shining and attractive finish.

10.4 The Colour of Semiconductors

In *semiconductors*, a narrow gap separates the *valence band* from the *conduction band*. The colour of a pure semiconductor is governed by this energy. When the energy gap is relatively large light photons are not energetic enough to excite an electron from the valence band to the conduction band and so are not absorbed. The material will appear transparent to these wavelengths. On the other hand, if the minimum energy is quite small and lies in the infrared region the semiconductor will absorb (and reflect) all of the visible spectrum and take on a metallic appearance. If the band gap falls in the visible, the semiconductor will absorb all photons with an energy greater than the band gap and not those with a smaller energy. This will cause the material to be strongly coloured.

For example, the pigment *vermilion*, which is produced from the mineral *cinnabar*, or mercuric sulphide (HgS), has a band gap of 3.2×10^{-19} J (2.0 eV). This energy corresponds to the red-orange region of the spectrum. All shorter wavelengths than this are associated with more energetic photons and these will be absorbed. The colour perceived will be due to the photons which are not absorbed, red and orange. Crystals of vermilion are shown in Plate 10.1. The pigment *cadmium yellow*, or cadmium sulphide (CdS), has a band gap of 4.16×10^{-19} J (2.6 eV), which corresponds to the green-blue part of the visible. Photons of lower energy, red, orange, yellow and some green, will not be absorbed while the higher energy greens and blue, indigo and violet will be. The net result is that the pigment appears yellow to the eye.

A less widely used pigment these days is *orpiment*, arsenic trisulphide, As_2S_3. The mineral name "orpiment" is a corruption of the Latin *auri pigmentum*, golden paint, and it is also known as the artist's colour King's yellow. It is readily prepared as a canary yellow precipitate when hydrogen sulphide gas is passed into solutions containing As^{3+} ions. The pigment has fallen into disfavour because of its toxicity and tendency to give off poisonous molecules when exposed to damp air. Indeed, almost all coloured sulphides, regardless of toxicity, have figured as artists pigments in one context or another in earlier centuries.

Interestingly, band gaps can be finely tuned by making solid solutions of semiconductors. This can be illustrated with respect to cadmium sulphide and the very similar cadmium selenide, CdS and $CdSe$. Both of these compounds have the same crystal structure, namely that of *wurtzite*, one of the forms of zinc sulphide, ZnS. CdS, as we have seen, absorbs high energy photons from violet to green-blue. $CdSe$ has a small band gap of 2.56

$\times 10^{-19}$ J (1.6 eV) and absorbs all the visible wavelengths. It appears black to the eye. The sulphur and selenium atoms in these two compounds are of a similar size, which allows one to replace the other readily. If a solid solution is made with a general formula $CdS_{1-x}Se_x$ the band gap gradually changes from that appropriate to CdS at x = 0 to that appropriate to CdSe at x = 1.0. At x = 0 the photons absorbed are only those in the violet to blue-green region of the visible but as x increases the range of absorbed photons moves towards red and infrared. The colour perceived gradually changes from yellow, at x = 0, to orange to red and ultimately to black as *x* increases. The material $CdS_{0.25}Se_{0.75}$ is the pigment cadmium orange.

Many metal sulphides show a similarity to metals, both visually and electronically, indicating that free electrons are present. The most widely known example of this similarity is found in the compound *pyrite*, FeS_2, also known as *fool's gold*. The physical properties are not entirely metallic, however, and pyrite is brittle rather than malleable, as is gold. Admixture of copper sulphide, CuS, with pyrite produces the mineral *chalcopyrite*, with a nominal formula $Cu_2Fe_2S_4$. This material also has a metallic appearance, and takes on a variety of golden or purplish hues depending upon the exact composition, as illustrated in Plate 10.2a,b.

The band gap of many semiconductors has been found to decrease slightly with temperature. Although this effect is small it can lead to interesting colour changes. White zinc oxide, ZnO, absorbs in the near ultraviolet. At high temperatures the decrease in the band gap means that some violet light is absorbed. The material will then become yellow to the eye. This effect is also noticeable with the yellow indium oxide In_2O_3. At room temperature this absorbs in the green-blue. As the temperature increases the absorption shifts towards the lower energy red, causing the oxide to take on a much deeper yellow-brown colour. The effect of colour variation with temperature is known as *thermochromism*. This term was encountered in Section 6.5 with respect to liquid crystal thermometers. (Note that the name thermochromism applies to the change of colour with temperature. That is, it does not describe the mechanism, only the observed effect. The two examples mentioned have quite different mechanisms.)

10.5 Impurity Colours in Insulators and Semiconductors

Due to a large band gap, most insulators are colourless. Impurities can change this state of affairs and we have already considered "coloured"

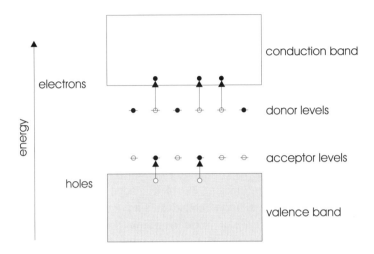

Figure 10.5 Acceptor levels, which can accept electrons from the valence band, and donor levels, which can donate electrons to the conduction band, in an insulator

transition metal ions or lanthanides in glasses and gemstones. Surprisingly there are two ways in which normally colourless materials can become coloured by the addition of *colourless* impurity atoms.

The first of these involves the creation of new energy levels within the band gap. This modification is represented on band diagrams in a similar way to that shown in Figure 10.5. The excitation of electrons to and from these levels will give rise to colours when the energy difference falls in the visible range. The impurities are classed as *donors* if they contribute electrons to the conduction band, or *acceptors* if they take up electrons from the valence band.

These processes are responsible for coloration of diamonds containing as little as 1 part in about 100 000 nitrogen or boron impurities. Diamond has a band gap (of about 8.65×10^{-19} J, 5.4 eV) which is too large to absorb visible light and diamonds are therefore clear. Nitrogen impurities occupy normal carbon atom positions. In so doing a band of energy levels 6.41×10^{-19} J (4.0 eV) below the conduction band is created. Nitrogen has five bonding electrons, one more than carbon. Four of the electrons around each impurity nitrogen atom are used to fulfil the local bonding requirements of the crystal structure and one electron remains unused. The extra electron, one per nitrogen atom impurity, sits in the energy gap to form a donor level.

The electrons within this new set of energy levels can be excited to the conduction band, as depicted in Figure 10.5; an energy of 6.41×10^{-19} J

(4.0 eV) being required. Although this energy is in the ultraviolet, and corresponds to a wavelength of 310 nm, the absorption spectrum is actually broad enough for the low energy tail to creep into the violet beyond 400 nm. Thus diamond containing a few atoms of nitrogen is found to absorb slightly in the violet region, giving the stones a slightly yellow aspect. As the nitrogen concentration increases the colour becomes greener.

In the case of boron impurities a complementary situation occurs. Each boron impurity atom occupies a carbon position, which results in the creation of a set of new energy levels just 0.64×10^{-19} J (0.4 eV) above the valence band. Boron has only three outer bonding electrons instead of the four found on carbon. The missing electron needed to fulfil the chemical bonding requirements in diamond constitutes an *acceptor level*, as drawn in Figure 10.5. The transition of an electron from the valence band to this acceptor level absorbs energy principally in the infrared but the high energy tail of the absorption band spills into the red at 700 nm. The boron-doped diamonds therefore absorb some red light and leave the gemstone with an overall blue colour.

10.6 THE F-CENTRE

Exposure of alkali halide crystals to high energy radiation such as X-rays causes them to become brightly coloured. In early studies the source of the colour was attributed to the formation of entities that were called *Farb-zentren*, or *colour centres*, now shortened to *F-centres*. One significant fact is that, regardless of the type of radiation used, the colour produced in any particular crystal is always the same. Thus, F-centres in sodium chloride, NaCl, are always an orange-brown colour. Measurement of the absorption spectra of these crystals reveals a more or less bell shaped curve of the type shown in Figure 10.6. Room temperature data for colour centres in alkali halide materials are collected in Table 10.2. The peak of the absorption curve, λ_{max}, is seen to move to higher wavelengths as both the alkali metal ion size and halide ion size increase. The appearance of the colour centre-containing crystal is the complementary colour to that removed by the absorption band.

There are, it turns out, other ways in which one can produce F-centres in alkali halide crystals apart from using ionising radiation. One of these involves heating the crystals at high temperatures in the vapour of an alkali metal. It is notable that the exact metal does not matter as long as it is an alkali metal. That is, if a crystal of potassium chloride, KCl, is heated in an

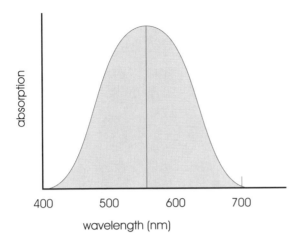

Figure 10.6 A typical bell-shaped absorption curve due to F-centres in potassium chloride, KCl. Curves for other F-centres are similar in shape but are displaced to other wavelengths

Table 10.2 Alkali metal halide F-centres

Compound	λ_{max} (nm)	Colour of crystal	Lattice parameter (nm)
LiF	235, UV	Colourless	0.4073
NaF	345, UV	Colourless	0.4620
KF	460, blue	Yellow-brown	0.5347
RbF	–	–	0.5640
LiCl	390, UV (just)	Yellow-green	0.51295
NaCl	460, blue	Yellow-brown	0.5641
KCl	565, green	Violet	0.6293
RbCl	620, orange	Blue-green	0.6581
LiBr	460, blue	Yellow-brown	0.5501
NaBr	540, green	Purple	0.5973
KBr	620, orange	Blue-green	0.6600
RbBr	690, red	Blue-green	0.6854

atmosphere of sodium vapour the typical violet KCl F-centres are formed, and not the yellow-brown NaCl colour centres. Another way of introducing F-centres into alkali halide crystals is to pass an electric current through heated samples and electrolyse them. In this case the typical F-centre colour is seen to move into the crystal from the cathode region. Once again, the colour depends upon the crystal being electrolysed and not the exact nature of the cathode.

These observations suggest that the centres are associated with defects in the crystal structure rather than the precise chemical elements which

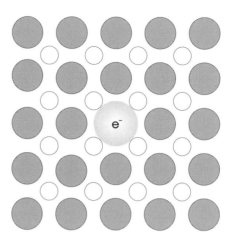

Figure 10.7 The structure of an F-centre in an alkali halide crystal. The anions are shown as filled circles and the cations as smaller open circles. The colour centre consists of an electron trapped at an anion vacancy

constitute the compound. The defect responsible turns out to be an anion vacancy plus a trapped electron, as illustrated in Figure 10.7. The trapping is due to the anion vacancy, which, being surrounded by six cations, can be thought of as bearing a positive charge. The electron in this location is able to absorb electromagnetic radiation and yield the type of absorption spectrum shown in Figure 10.6.

An examination of the data in Table 10.2 suggests that there is a correlation between the position of the absorption maximum, λ_{max}, and the chemical constitution of the alkali halide. For example, the colour centres in KF, NaCl and LiBr all have identical values for λ_{max}. This can be explored if we suppose that the compounds can be treated as ionic materials, a model which works quite well for the alkali halides.

Let us suppose that the energy which corresponds to λ_{max} can be equated to the energy to remove the trapped electron completely from the vacancy. The potential of this site can be assessed using an ionic model and adding the contributions of all of the other ions in the crystal together. The result is called the *Madelung potential* of the site. For the alkali halides it is given by:

$$\phi_i = -e\alpha_i \,/\, 4\pi\varepsilon_0 r_{AB}$$

where ϕ_i represents the Madelung potential at the site occupied by an ion of type i, which has a charge of e, α_i is a constant which lies close to 1, ε_0 is

the permittivity of free space and r_{AB} is the separation of the ions A and B
in the crystal of formula AB. Setting α_i equal to 1.0, the energy to remove a
trapped electron from this site, E, is:

$$E = -\phi_i \, e$$
$$= e^2 \, / \, 4\pi\varepsilon_0 r_{AB}$$

It is of interest to evaluate this. Taking NaCl as a test case and equating half
of the unit cell edge to the distance r_{AB} (0.282×10^{-9} nm for NaCl) it is
found that the energy, E, is 8.18×10^{-19} J. Using the equation:

$$E = hc \, / \, \lambda_{max}$$

in which h is Planck's constant and c is the velocity of light in a vacuum,
the wavelength of the emission peak, λ_{max}, is predicted to occur at 243 nm.
Reference to Table 10.2 shows that this figure is considerably too low.

One way in which it is possible to modify the calculation is to assume
that the electron in the F-centre is shielded from the full effect of the site
potential by a "screening factor". The easiest screening factor to employ is
the relative permittivity of the bulk solid, ε_r. Within this approximation the
energy required to release the electron from an F-centre in NaCl will be
given by:

$$E = e^2 \, / \, 4\pi\varepsilon_0\varepsilon_r r_{AB} \qquad (10.1)$$

where ε_r represents the relative permittivity of NaCl, which has a value of
5.1. Repeating the calculation gives a value of 1.60×10^{-19} J for the energy
and a value of 1239 nm for the absorption peak λ_{max}. This value is clearly
too high and the degree of shielding has been overestimated.

Nevertheless, the principle of the method seems reasonable and it may
be better to write:

$$E = K \, / \, r_{AB}$$

where K is $e^2/4\pi\varepsilon_0 s$ and s is an empirically determined screening factor.
Moreover the energy is related to the absorption peak wavelength by:

$$E = hc \, / \, \lambda_{max}$$

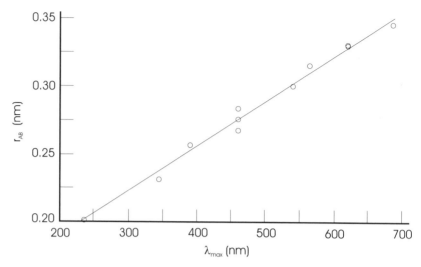

Figure 10.8 The absorption peaks, λ_{max}, for alkali halide colour centres, plotted against r_{AB}, equal to half the lattice parameter of the host crystal

Setting these two equations equal and rearranging slightly gives:

$$\lambda_{max} = (hc\ /\ K)\ r_{AB}$$

A graph of λ_{max} against r_{AB} should be a straight line with a slope of $(hc\ /\ K)$. The result is shown in Figure 10.8, where the values of r_{AB} are computed from half of the lattice parameters of the alkali halides, listed in Table 10.2.

It is seen that a linear relationship is obtained. Using this graph and extrapolating it is possible to estimate the value of λ_{max} for RbF (not included in the table) of 515 nm. Furthermore, an estimate of the value of K, taken from the slope of the graph suggests a value for the screening constant appropriate to NaCl is 2.35.

10.7 ELECTRON AND HOLE CENTRES

Since the original studies of F-centres the term colour centre has broadened in meaning to include any crystal defects with associated trapped electrons or holes. These are called *electron excess* or *hole excess* centres, respectively.

The F-centre is an *electron excess centre* and arises because the crystal contains a small excess of metal. Similar metal excess F-centres exist in compounds other than the alkali halides. An example is provided by the mineral *Blue John*. This is a rare, naturally occurring form of fluorite, CaF_2. (The name "Blue John" is a corruption of the French term "bleu-jeune" which was used to describe the blue form of the normally yellowish fluorite crystals found in nature.) The coloration is caused by electron excess F-centres identical to those just described except that the anion vacancy is surrounded by four cations instead of six as in the alkali halides. It is believed that the colour centres in Blue John were formed when the fluorite crystals were fortuitously located near to uranium compounds in the rock strata. Radioactive decay of the uranium produced the energetic radiation necessary to form colour centres.

One of the best understood hole excess centres gives rise to the colour in smoky quartz and some forms of amethyst. These minerals are essentially crystals of silica, SiO_2, which contain small amounts of either Al^{3+} or Fe^{3+} as impurities. Consider the effect of Al^{3+}. The trivalent ion substitutes for tetravalent silicon in the structure, and charge neutrality is preserved by way of incorporated hydrogen. This is present as H^+ in exactly the same amount as the Al^{3+}. The colour centre giving rise to the smoky colour is formed when an electron is liberated from an $[AlO_4]^{-5}$ group by ionising radiation and is trapped on one of the H^+ ions present. The reaction can be written as:

$$[AlO_4]^{5-} + H^+ \rightarrow [AlO_4]^{4-} + H$$

It is seen that the $[AlO_4]^{4-}$ group is now electron deficient and can be thought of as $[AlO_4]^{5-}$ together with a trapped hole. The colour arises when the trapped hole absorbs radiation in precisely the same way as with the boron impurity in diamond.

The situation with Fe^{3+} impurities is similar. Once again the impurity ion substitutes for tetravalent Si^{4+} in the quartz. These crystals are now a pale yellow colour due to the crystal field splitting of the d-electron levels on the Fe^{3+} ions, as described in Chapter 7. In this form natural crystals are known as *citrine*, a semiprecious gemstone. On irradiation, $[FeO_4]^{4-}$ groups form by interaction with H^+ ions, as described by the equations for $[AlO_4]^{4-}$ above. The colour centre, which can be thought of as an $[FeO_4]^{5-}$ group plus a trapped hole, is able to absorb light, giving the crystals the purple amethyst coloration.

Although the exact cause of the coloration is not completely clarified, the rutile form of the white pigment titanium dioxide, TiO_2, seems to be coloured by hole centres formed as a consequence of the incorporation of colourless Ga^{3+} ions. When crystals of rutile are heated with gallium oxide, Ga_2O_3, small quantities of Ga^{3+} impurity are readily incorporated into the structure. Initially clear single crystals of rutile take on a yellow-orange colour, illustrated in Plate 10.3. The impurity Ga^{3+} ions enter the rutile structure and substitute for Ti^{4+} ions in octahedral sites. The sites occupied by these impurities are surrounded by six anions and each can be considered to have an overall negative charge, allowing it to trap a positively charged hole. The liberation of the holes absorbs energy towards the violet end of the spectrum and colours the crystals yellow-orange.

We close by noting that many porcelain objects such as sinks which were in buildings near to the atomic power station disaster at Chernobyl became blue after the accident. The colour was due to the formation of colour centres produced as a result of exposure to large doses of ionising radiation. This explanation is similar to that invoked for smoky quartz, given above, as the sinks were constructed from ceramics containing both aluminosilicate and H^+ ions. However, precise details have yet to be given for any of these rather complex materials.

10.8 Answers to Introductory Questions

Why are metals shiny in bulk, but black as powders?
Metals are shiny when in the form of fairly flat polished surfaces. The reason is that a metal has a large absorption coefficient and is able to absorb all of the incident photons in its surface region. These photons, however, are all re-emitted, causing the material to look reflective. If the surface is rough, especially when the metal is powdered, the re-emitted photons are re-absorbed by neighbouring grains of metal and the whole looks dark. However, if the metal is hammered into flakes and these are well separated, they again can look shiny. This effect is used with flakes of aluminium in some "metallic" car paints.

Why are gold and copper coloured?
The reflectivity of gold and copper is highest at the red end of the visible and so the colour reflected by these metals is dominated by reds and yellows. The reflectivity is controlled by the band structure of these metals and the probability of a photon being able to excite an electron to higher energy levels in the solid. For most metals this is constant across the energies of the visible, but for copper and gold the change is rather large.

What produces the intense colours in the paints vermilion and cadmium yellow?

The paint pigment vermilion is the sulphide HgS and the paint pigment cadmium yellow is the sulphide CdS. These are semiconducting materials and the band gap energy corresponds to the energy of one wavelength of visible light, λ_c (vermilion) or λ_c (cadmium yellow). Each sulphide will absorb all wavelengths of light shorter than the respective λ_c values. The colours seen are produced by subtractive coloration.

This interpretation can be confirmed by looking up the values for the band gaps of these phases. Using these to find the value of the wavelength λ_c we find:

HgS: λ_c (vermilion) = 590 nm (yellow – orange)

CdS: λ_c (cadmium sulphide) = 512 nm (green)

Thus HgS absorbs every wavelength shorter than yellow-orange, leaving red to be reflected to the observer. CdS absorbs all radiation with a wavelength shorter than green, leaving yellow and red. The eye is most sensitive to yellow and this is the colour perceived.

How can colourless impurities turn diamonds blue or yellow?

Diamonds are insulators coloured by impurities which introduce new energy levels into the band gap of the pure material. Nitrogen introduces donor levels which absorb in the ultraviolet, just tailing into the visible at violet. This colours the diamonds yellow by subtractive coloration. Boron introduces acceptor levels which absorb in the infrared, just tailing into the red. Thus boron impurities colour the diamonds blue by subtractive coloration.

In what way do holes change quartz into amethyst?

Amethyst consists of a crystalline form of silica, SiO_2, containing a small amount of Fe^{3+} together with H^+. The Fe^{3+} sits in a tetrahedron of oxygen ions and captures the charge of a proton to form an $[FeO_4]^{4-}$ group and a neutral H atom. The $[FeO_4]^{4-}$ group can be thought of as having has an electron deficiency (a trapped hole) which can absorb radiation. This absorption produces the purple colour of amethyst. The $[FeO_4]^{4-}$ group is thus a colour centre.

10.9 FURTHER READING

The topics covered in this chapter are discussed in:

K. Nassau, The Physics and Chemistry of Colour, Wiley-Interscience, New York (1983), Chs 8, 9.

More information on band theory and related topics will be found in:

P. A. Cox, The Electronic Structure and Chemistry of Solids, Oxford University Press, Oxford (1987).
L. Solymar and D. Walsh, Lectures on the Electrical Properties of Materials, 5th edition, Oxford University Press, Oxford (1993).

Colour centres are described in:

R. J. D. Tilley, Principles and Applications of Chemical Defects, Stanley Thornes, Cheltenham (1998), Ch. 7.
W. B. Fowler, Colour Centres in Ionic Crystals, in Treatise on Solid State Chemistry, Vol. 2, Defects in Solids, ed. N. B. Hannay, Plenum, New York (1975).

The Madelung potential and related matters are described by:

M. O'Keeffe, in Structure and Bonding in Crystals, eds M. O' Keeffe and A. Navrotsky, Vol. 1, Chapter 13, Academic Press, New York (1981).

10.10 PROBLEMS AND EXERCISES

1. Calculate the reflectivity of mercury, chromium and titanium for yellow light using the following data, which is appropriate to that colour:

$$Hg \quad n = 1.620, \quad k = 4.751$$
$$Cr \quad n = 3.18, \quad k = 4.41$$
$$Ti \quad n = 1.92, \quad k = 2.67.$$

2. Considering the information presented in Figure 10.5, suggest the best materials for mirrors operating in the near infrared and near ultra-violet.

3. Diamond, silicon, germanium and α-tin all have the same structure. The colour, however, varies widely, from transparent in the case of diamond, to steel-grey for silicon, lustrous grey-white for germanium and pale shining grey-white for α-tin. Explain the appearance of these materials.

4. The compounds ZnS, ZnSe and ZnTe all have the same (sphalerite) structure and the following band gaps:

$$
\begin{array}{ll}
\text{ZnS} & 3.54 \text{ eV} \\
\text{ZnSe} & 2.58 \text{ eV} \\
\text{ZnTe} & 2.26 \text{ eV}
\end{array}
$$

What colour will they appear to be? If solid solutions of the type ZnS_xSe_{1-x} are made what colours would these show as x varies?

5. How will applied pressure change the appearance of a coloured semi-conductor such as CdS, ZnSe or HgS?

6. On irradiation, a crystal of composition $Na_{1-x}K_xBr$ is found to incorporate colour centres showing a value of the absorption peak, λ_{max}, of 615 nm. Assuming that there is a linear relationship between crystal composition and lattice parameter (in accordance with Vegard's law) estimate the composition of the crystal.

7. Explain how the structure of an F-centre in Blue John differs from that in potassium fluoride (KF). How might the differences affect the electric field experienced by the trapped electron?

CHAPTER 11

FIBRE OPTICS AND DATA TRANSMISSION

Why was it necessary to change to optical
communication links?
What technology was needed to set up optical links?
What are the chemical and physical restrictions
on optical links?
What future developments in optical links might
be anticipated?
How can doped fibres be made to amplify optical signals?

11.1 FIBRE OPTICS

It was Tyndall, in 1870, who first showed that light could be transmitted
along a jet of water even if the path was curved. The reason for the transmission is that total internal reflection at the water–air interface, described
in Chapter 2, prevents the light from escaping. Shortly after this, the
transmission of light within a glass rod was also demonstrated. As glass can
be easily dawn into fine fibres it soon became clear that bundles of fibres
could be used for the remote illumination and viewing of inaccessible or
dangerous areas. The subject of light transmission along thin fibres of glass,
plastic or other transparent materials is frequently referred to as *fibre optics*.

Until the mid-1950s fibre optics remained something of a curiosity. This
was because the transparency of the glass was not very good, due to the
presence of transition metal ions and other impurities. Moreover, different
colours tended to separate because of the dispersion of the refractive index
of the glass, resulting in strong chromatic aberration and producing
unsatisfactory images. Despite these problems, use was made of short
lengths of glass fibres for decorative purposes and lighting. However, during
these years the main use of glass fibre bundles was in medicine, making the
examination of internal organs possible without surgery.

The situation changed in the mid-1960s. The impetus was for rapid
communication of large amounts of data along secure lines and for this

glass fibres were deemed ideal. As a spin-off from developments in communication technology, uses of fibre bundles as long distance light guides, for remote viewing of inaccessible objects or dangerous devices, in medical imaging and in many applications of laser technology has burgeoned. However, the developments which led to these changes is best described with respect to the development of fibre optic communications. This technological advance, embodying as it does the vast data flows characteristic of the Internet, has already had an enormous impact upon the day to day lives of many millions of people and will certainly mark a turning point in the history of mankind.

At present, optical fibre communications use infrared wavelengths. The principles explained in earlier chapters, though, will be found to be vital in understanding the development of the systems presently in use and the subject is of such importance and interest that it would be unfortunate not to include some discussion in this book. Moreover, although there are problems associated with the move to visible wavelengths, there are also advantages to be gained by this shift. Future developments suggest that the move towards shorter wavelengths will continue and visible light pulses of many colours may one day be the favoured method of high density data transmission.

11.2 HISTORICAL BACKGROUND TO OPTICAL COMMUNICATIONS

Communications are central to human activity. As a result of this, there has been, for a long time and especially during times of war, a need for rapid long distance communications. *Signal fires*, such as those used in Britain at the time of the Spanish Armada, are effective, but they are limited to a predetermined message. In France, Chappe developed an extensive *semaphore* system that operated successfully between 1790 and 1880. It could transmit a message 200 km in 15 min. This system was only superseded in 1880 by the *electric telegraph*, using the *Morse code*. This was itself superseded by "*wireless telegraphy*" using radio waves, and then by the telephone.

It became apparent in the 1960s that the current system of communication was becoming saturated. There was an increasing desire for higher volumes of data transmission and transmission over longer distances. The drive behind this was mainly commercial and social rather than military. For example, more and more people wanted telephones. At the time most telephone messages were carried by copper cables which were accom-

Table 11.1 Early history of optical communications

Year	Development
1870	Tyndall showed that light could be transmitted down a water jet
1880	Bell invented the photophone. Voice vibrated a thin mirror which modulated a reflected beam of sunlight directed onto a selenium detector
1927	Baird patented the idea of light transmission through glass with respect to colour television. The idea failed due to poor quality glass fibres
1950s	Low grade optical fibres used in medicine
1960	Ruby laser invented
1966	Kao and Hockham proposed optical communications if pure glass could be made
1970	The first pure fibre produced by Corning glass
	The first diode laser fabricated
1976	The first trial optical communication system installed

modated in ducts below the streets in major cities. The ducts were full and expansion of the existing system was not regarded favourably. Can you imagine the disruption that would be caused by digging up all roads in major cities to install new ducting? Apart from that, the process would be slow and costly. How long would people in outlying districts be prepared to wait for the new lines?

How could the system be improved? From several possible data transmission systems envisaged, one using light as the carrier was chosen. The idea was not altogether new, as the information in Table 11.1 reveals. However, light communications had many potential advantages if the technological difficulties that had dogged the earlier attempts could be overcome. Not least was the fact that very small diameter optical cables could carry far more information than bulky copper cables, allowing the existing understreet ducting to be used. (Compare, e.g. TAT-7 and TAT-8 in Table 11.2.) We now know that an optical system was chosen and is in place and it is true to say that the use of optical fibre communications has resulted in a revolution in data transmission. Some milestones in the switch to optical fibres are listed in Table 11.2. The rest of this chapter explains some of the fascinating work that made this transformation possible.

11.3 THE FUNDAMENTALS OF FIBRE-OPTIC COMMUNICATIONS LINKS

Data are carried in optical communications by a series of pulses of light encoded so that information can be stored and retrieved. Much information,

Table 11.2 Long-distance communications cable development

Year	Diameter (cm)	Repeater spacing (km)	No. of voice circuits	Development
1927				27 January: first public telephone service across the Atlantic begins
1951	1.6	70.5	36	TAT-1 (Trans Atlantic Telephone) copper cable laid
1959–1976	5.3	9.5	4200	Installation of further transatlantic copper cables. The specifications shown are for TAT-6 (1969)
1977	–	–	–	First optical fibre links installed in the North American phone network. First unrepeatered 140 Mbit, 9 km, optical fibre system installed, Hitchin–Stevenage, UK
1980	–	–	–	STC installed world's first experimental submarine optical fibre cable in Loch Fyne, Scotland. First UK optical fibre system installed, 140 Mbit, London–Basildon
1983	5.3	9.5	4200	TAT-7 copper cable laid. UK first 140 Mbit monomode optical fibre system, 27 km, Luton–Milton Keynes
1984	–	–	–	ANZCAN, probably the last copper submarine cable, linking Australia, New Zealand, Hawaii and Canada. UK first operational submarine optical cable, 8 km, Portsmouth–Isle of Wight
1985	–	–	–	Optical fibre communication equipment sales reach $1 billion. STC install optical cable linking Broadstairs, Kent–Ostend, Belgium, 112 km
1988	2.1	50	40 000	TAT-8, optical cable, running from USA, to the Bay of Biscay, where it splits to Western France and the West of England. TRANSPAC-3, HAWAII-4, optical links between continental USA and Hawaii, Guam and Japan. Capacity of 38 000 voice circuits Lasers are developed allowing 10^6 conversations or 10^4 TV channels to be transmitted simultaneously down a single fibre
1990–5	–	–	–	Many TV and voice channels proposed and installed using optical links

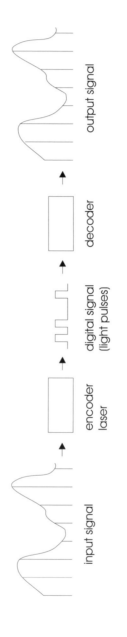

Figure 11.1 The amplitude of an in input signal (far left) is measured at set intervals and transcribed into light pulses as a digital signal (centre) by an encoder and laser. These are transmitted via an optical fibre to a decoder which recreates the original signal form (far right)

air

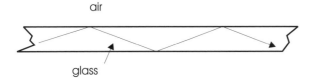

glass

Figure 11.2 A ray of light trapped in a fibre by total internal reflection

such as that of a voice, is an irregular waveform, as shown in Figure 11.1. This irregular signal is converted into digital form by measuring the amplitude of the signal at regular intervals. These amplitudes are measured over a scale of 0 to 256 (2^8) and each amplitude is then represented in binary code by a sequence of 8 bits (i.e. a sequence of eight zeros and ones) in an encoder. These are turned into a sequence of light pulses given out by a semiconductor diode laser (see Chapter 13) so that a pulse, for example, could represent a 1 and no pulse a 0.

To make a workable system a number of requirements are necessary. The foremost of these is a light source that is intense. Further thought reveals that it must be nearly monochromatic and must be capable of being switched on and off at enormous speeds. It is equally essential to construct a transparent optical wave carrier. (Open atmosphere transmission is out of the question since it is essentially restricted to line of sight.) There are a number of choices regarding the best fibre material to use. Polymers would seem to be a good option, but unfortunately these materials absorb too strongly to allow long distance operation. The best material for a signal carrier was found to be silica (SiO_2) glass. This is still the principal material in use because it is strong, easy to fabricate into fibres, plentiful and inexpensive.

The light pulses launched into the fibre are constrained to stay within the fibre by total internal reflection as shown in Figure 11.2. An essential requirement for total internal reflection is that the external medium must possess a lower refractive index than the fibre. The glass surface is easily damaged, which will seriously degrade performance and so a successful fibre needs a *surface cladding* of lower refractive index compared to the core of the fibre. The present day fibres consist of a higher refractive index glass core and a lower refractive index glass cladding. The core and the cladding make up a single glass fibre.

To complete the system, an input device to encode the original signal and output equipment to decode it are required. To end up with a reasonable representation of the input signal it is necessary to measure the wave amplitude at least twice the highest frequency in the wave. Taking a voice signal as typical, the highest frequency is about 4000 Hz so that the signal

Table 11.3 Evolution of optical-fibre communications

Year	Development
May 1977	First commercial evaluation in Chicago. Twenty-four fibres, each of which could carry 44.7 Mbit s^{-1} (672 one-way voice channels). An LED light source with a wavelength of 0.87 μm was used
1978	10 Gbit km s^{-1} (10^{10} bits) transmission rate achieved with 0.87 μm LED and graded index fibres
1979	Silica fibres reach intrinsic attenuation limitations of 0.2 dB km^{-1} at 1.5 μm wavelength (although this wavelength is still not available for use yet)
1980	100 Gbit km s^{-1} transmission rate achieved, by a change to 1.3 μm wavelength radiation and the first monomode fibres
1982	Reduction in the hydroxyl content of fibres gives enhanced performance in the 1.2–1.6 μm wavelength range. Monochromatic laser diodes emitting at 1.55 μm wavelength available. Transmission rates of 100s of Gbit km s^{-1} achieved
1989	Erbium fibre amplifiers using 1.48 and 0.98 μm pump radiation perfected

must be measured about 8000 times each second. As each amplitude is recorded as 8 bits we need about 64 000 bits per second to be transmitted. At its simplest level it is necessary to turn the laser on and off at this frequency to send one voice signal.

After transmission, a sensitive light detector at the receiving end decodes the light pulses into amplitude values. For a voice signal these are used to recreate the waveform with a speaker. Finally, because the optical signal will tend to degrade or diminish over distance it is necessary to have some sort of signal regenerators (repeaters or amplifiers) installed along the optical line.

The way in which these requirements have been welded into highly successful communications systems over the last 20 years can be seen from Table 11.3.

11.4 ATTENUATION IN GLASS FIBRES

Attenuation describes the loss of light intensity as the signal is transmitted along the fibre. The units of attenuation are dB km^{-1}. (The bel (B) is a measure of power level used especially with sound. As it is a large quantity the more usual unit is the decibel (dB) which is 0.1 bel. An amount of power P (W m^{-2}) has a power level in decibels of [10 log(P/P$_0$)] where P$_0$ is a reference value.)

$$\text{Attenuation} = -10 \log_{10} [P(x) / P(0)] / x$$

where P(0) is the power at x = 0, P(x) is the power x kilometres along the cable and x is the length of the link in km. For a material showing an attenuation of 1dB km^{-1} an input power of 10 W would give an output power of 7.9 W after 1 km. Ordinary window glass has an attenuation of about 10 000 dB km^{-1}. The *spectral response* of a fibre defines the way in which the fibre attenuation changes with the frequency of the radiation being transmitted.

Attenuation is caused by a combination of absorption and scattering within the glass. *Extrinsic* attenuation is caused by poor processing or fabrication techniques, and may be due to artefacts such as bubbles, particles, impurities and variable fibre dimensions. In modern optical fibre manufacture these have been eliminated and no longer pose a problem. *Intrinsic* attenuation is a property of the pure material itself, and cannot be removed by processing. It is the ultimate limit on the performance of the material and arises mainly from two factors, Rayleigh scattering and absorption due to lattice vibrations, mostly in the infrared region. Absorption due to electronic transitions, mostly at high energies and associated with ultraviolet wavelengths, do not figure significantly in present day applications, but may become important if shorter signal wavelengths are contemplated in the future.

Rayleigh scattering arises from small inhomogeneities in the glass which cause changes in refractive index. This variation is an inevitable feature of the non-crystalline state and cannot be removed by processing. As Rayleigh scattering is proportional to λ^{-4} where λ is the wavelength of the optical pulse, the effect is more important if short wavelengths are used. For any particular glass most of the factors affecting Rayleigh scattering are constant and cannot be easily changed. However, materials with a low refractive index and glass transition temperature[6] tend to exhibit low Rayleigh scattering. Infrared absorption, also referred to as *phonon absorption*, occurs when the lattice vibrations of the solid match the energy of the radiation. This converts the signal energy into heat. It is a function of the mass of the atoms in the glass and the strength of the chemical bonds between them and results in a decrease in the transparency of the glass at long wavelengths. The dependence on wavelength of absorption due to electronic transitions can often be expressed by a formula of the type:

---

[6] The glass transition temperature of a glass is the temperature at which the material can be considered to be a solid rather than a liquid. It is a widely used parameter in glass manufacture and processing.

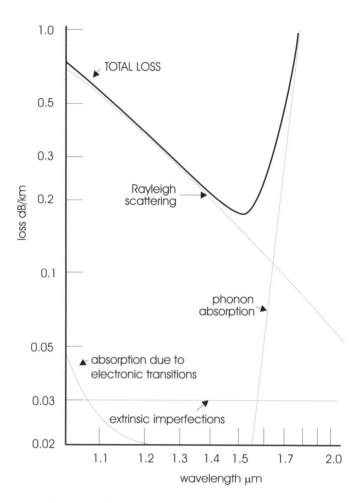

Figure 11.3 Schematic illustration of the contribution of Rayleigh scattering, phonon absorption at infrared wavelengths, electronic absorption at the shortest wavelengths, and extrinsic imperfections on the attenuation of a silica glass fibre

$$\text{electronic absorption} = B_1 \exp (B_2 / \lambda)$$

where B_1 and B_2 are constants relating to the glass used and λ is the wavelength of the radiation.

The effect of these intrinsic attenuation factors is shown in Figure 11.3. Remarkably, by 1979 the best silica fibres showed only intrinsic attenuation and had a loss of about 0.2 dB km^{-1} at 1.5 μm wavelength. This is currently the industry standard.

Despite this achievement, new fibre materials are constantly being explored. As was noted in Chapter 8 with respect to water, the absorption maxima caused by lattice vibrations can be manipulated both by changing the strength of the chemical bonds between the components and by changing the mass of the atoms linked. For example, silica (SiO_2) with rather light atoms and strong bonds, transmits satisfactorily only to about 2.5 μm and strongly absorbs radiation in the 8–15 μm wavelength range. One group of recently developed materials with great promise are glasses primarily made from fluorides of zirconium, ZrF_4, barium, BaF_2 and lanthanum, LaF_3, called *ZBLAN glasses*. Although the chemical bonding in fluorides is regarded as being as strong as in oxides, the heavy atoms move the absorption maximum to wavelengths of 17–25 μm. In addition they are found to have an attenuation which is only one hundredth of that of silica. They are, therefore, enormously attractive for long-distance high data density communications. More desirable characteristics are found in the arsenic triselenide (As_2Se_3) glasses, composed of very heavy atoms and linked by weak bonds. These do not absorb strongly until 44–46 μm and so show great potential for the transmission of infrared radiation. Unfortunately the chemical difficulties associated with making fluoride and selenide glasses have not yet been solved and they are not currently used in long-distance commercial applications.

11.5 CHEMICAL IMPURITIES

The preparation of high purity glass was one of the most important of the technologies needed to allow fibre-optic communications to become a reality. The enormous strides in improvement of glass purity which have been made since the earliest times is illustrated in Figure 11.4. In the original glass fibres, transition metal impurities caused difficulties because they absorb strongly in the visible. The gravest problem was iron, present as Fe^{2+}, and it is this ion which gives window glass its greenish tint. Even as low a concentration as 1 part per million of iron can result in an attenuation of 15 dB km^{-1}. The presence of transition metal cations was overcome by the preparation of silica using very high purity chemicals made available by the semiconductor industry. At present it is possible to purchase silica with no significant transition metal ion impurities present.

The most important impurity in silica fibres today remains hydroxyl (–OH) as depicted in Figure 11.5. Hydroxyl arises from water or hydrogen incorporation into the glass during fabrication. Flames used to melt silica

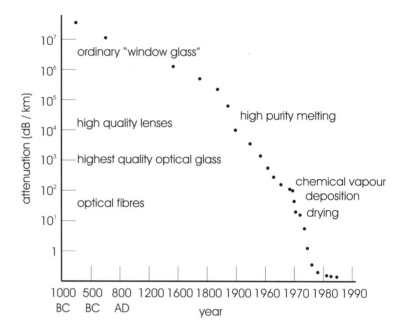

Figure 11.4 The improvements in the quality of glass through the ages. Recent improvements have been in response to the needs of optical fibre manufacturers. Currently silica fibres are routinely made with an attenuation of 0.2 dB km^{-1}

are rich in both of these impurities and any silica melted in a gas flame will be heavily contaminated by hydroxyl. An impurity level of 1 part per million can give an attenuation of 10^4 dB km^{-1} at a 1.4 μm signal wavelength. It is clear, therefore, that silica for optical fibre use must be melted in electric furnaces in a dry atmosphere to eliminate hydroxyl as much as possible. Despite careful processing, fibres currently in production still contain significant amounts of this impurity and hydroxyl remains an important source of attenuation.

11.6 Dispersion and Optical Fibre Design

Light is launched into a fibre as a short pulse. This pulse will spread out for several reasons as outlined below and illustrated schematically in Figure 11.6. This spreading is called *dispersion*. When dispersion was discussed in Chapter 2 it was defined in terms of the change of refractive index with

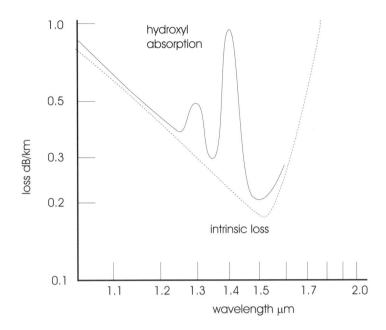

Figure 11.5 The contribution to fibre attenuation of hydroxyl (-OH) in the silica. The dotted curve is the total intrinsic loss curve shown in Figure 11.3

wavelength. In optical fibres, the *dispersion* is defined as the delay between the arrival time of the start of a light pulse and its finish time relative to that of the initial pulse. It is measured at half peak amplitude, as illustrated in Figure 11.7. Referring to this diagram, if the initial pulse has a spread of t_i seconds at 50 per cent amplitude and the final pulse a spread of t_f seconds at 50 per cent amplitude after having travelled d kilometres, the dispersion is given by

$$\text{dispersion} = (t_f - t_i) / d$$

The units are ns km^{-1}.

Dispersion will obviously arise if the light used is not monochromatic. An initially sharp pulse consisting of a group of wavelengths will spread out as it travels down the fibre as the refractive index depends on wavelength, and thus the light will travel at different speeds at different wavelengths. This effect is known as *wavelength dispersion*.

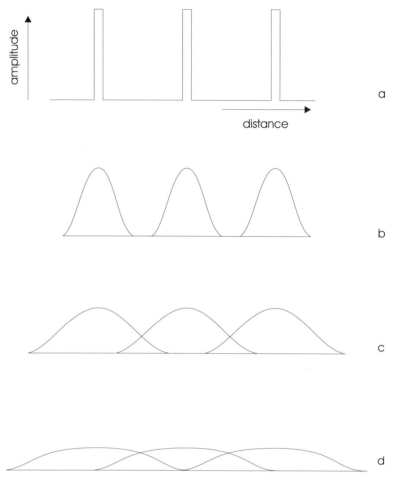

Figure 11.6 The gradual dispersion (or spreading) of a series of initially sharp light pulses (a), as they move along an optical fibre (b, c). At some point (d), the signal becomes lost

The first optical fibre communications installations used light emitting diodes (LEDs) (see Chapter 12). These had a spectral width of about 35 nm centred upon a wavelength of 0.82 μm. An instantaneous pulse of this radiation would spread to a width of about 65 cm after travelling 1 km along a silica fibre due to wavelength dispersion. This spreading sets a limit to the pulse frequency that can be employed and this in itself then limits the rate of data transmission. To illustrate this aspect of the problem, suppose that we arbitrarily assume that the pulse spreading after 1 km must amount to

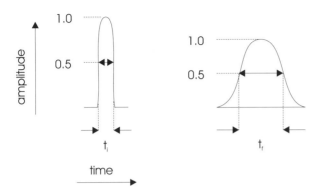

Figure 11.7 The dispersion of a light pulse is given by (t_f − t_i), the 50 per cent amplitude peak widths after the pulse has travelled 1 km

no more than half the separation of the initial pulses if we are still able to decode the signal. For the spread of 65 cm quoted, the initial pulses must be separated by at least 130 cm. This limits the frequency of the signal pulses to about 1.5×10^8 per second.

It is instructive to calculate the magnitude of wavelength dispersion for a number of hypothetical initial conditions in order to gain an understanding of how the system might be improved. For example, if a laser with a spectral range of 2 nm is substituted for the LED, the spreading of the initial sharp pulse will be only about 4 cm after 1 km. This means that the initial pulses can now be 8 cm apart, which will allow use of an increased pulse rate of 3×10^9 per second. In rough terms it is found that decreasing the spectral spread of the light source by a factor of ten leads to an increase in possible pulse rate by a factor of ten and gives a graphic illustration of why lasers are needed for rapid data transmission. This information is summarised in Table 11.4.

Unfortunately, even with completely monochromatic light, pulse spreading can still occur, due to the fact that the radiation can take various paths (called *modes*) through the fibre. A ray that travels along the axis of a fibre will travel less far than one which is reflected many times on its journey. The resultant pulse broadening is called *modal (or intermodal) dispersion*.

In order to overcome modal dispersion a number of different fibre types have evolved, and are illustrated in Figure 11.8. The earliest fibres were called *stepped index multimode* fibres. The step refers to a an abrupt change in the refractive index of the fibre, as in Figure 11.8a. These fibres have a large core region, making many modes possible. The ray in Figure

Table 11.4 Dispersion characteristics of optical fibres

Fibre type/ source	Wavelength spread (nm)^c	Dispersion (ns km^{-1})	Pulse width after 1 km (m)	Initial pulse separation (m)	Maximum pulse rate (s^{-1})
Stepped index	–	25	5.0	10	2×10^7
Graded index	–	1	0.2	0.4	5×10^8
LED	35	3.25	0.65	1.30	1.5×10^8
Laser	2	0.2	0.04	0.08	2.5×10^9

11.8 labelled h is known as a *high order mode* awhile the ray l is a *low order mode*. Stepped index multimode fibres are fairly easy to make and join, but have a lower performance compared to those described below.

It is useful to make trial calculations of the effects of modal dispersion and compare the results with those found for wavelength dispersion. For example, it has been estimated that for a stepped index fibre the difference in arrival time between the leading edge and trailing edge of an infinitely sharp initial pulse would amount to about 25 ns km^{-1}. This corresponds to a pulse width of 500 cm after 1 km. This means that initial pulses must be separated by 1000 cm. The limitation that this imposes upon pulse frequency is considerable, and about 2×10^7 pulses per second is the maximum that can be employed, as Table 11.4 shows. Because of this, stepped index fibres are adequate for short distance communications but not for medium or long distance links.

It is clear from this estimate that a greater limitation on data transmission rate in the first systems used was imposed by the refractive index profile of the fibre than by the light source characteristics. For this reason fibre refractive index profiles were improved before it was felt necessary to turn from LEDs to lasers, as Table 11.3 reveals.

The first advance on stepped index fibres was the *graded index* fibre. In this design, the refractive index of the fibre varies smoothly from high at the centre to low at the periphery of the core region. The refractive index gradient means that light travels faster and faster as it approaches the edge regions of the fibre. The velocity of mode a in Figure 11.8b will vary smoothly from lowest at the fibre centre to greatest near to the fibre edge. The velocity of mode b will be fairly constant and lower on average than mode a. The differences in path travel time between high order and low order modes is thus minimised by this velocity variation. Graded index fibres reduce modal dispersion by a factor of about 25 to about 1 ns km^{-1}. This amounts to a pulse width of about 20 cm after travelling 1 km so that

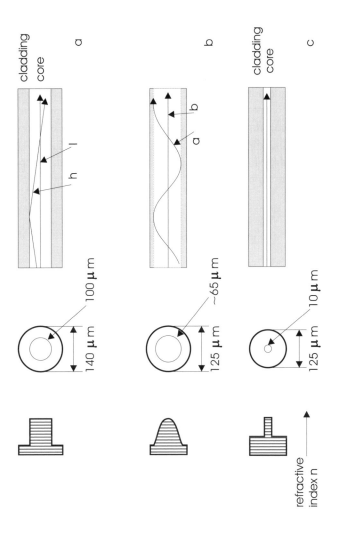

Figure 11.8 The three commonly used fibre types. (a) A stepped index multimode fibre. The path h represents a high order mode and the path l a low order mode. (b) A graded index fibre. The velocity of the ray b will be fairly constant while that of a will vary and increase towards the surface of the fibre. (c) A monomode fibre

an initial pulse rate of about 5×10^8 per second could be tolerated. These values are given in Table 11.4.

Even this improvement is insufficient for long distance communications. For best results *monomode fibres*, illustrated in Figure 11.8c, are now used. The number of possible modes is reduced simply by reducing the diameter of the core. When the core diameter reaches 10 μm or less only one mode can propagate and in principle modal dispersion is zero for these fibres. Monomode fibres therefore have a high performance but are harder to make and join.

At present all fibres in commercial use are made of silica, SiO_2. Although this is the best material available so far, it is not perfect. The dispersion is lowest at 1.3 μm but the minimum attenuation occurs at 1.5 μm, leading to some sacrifice of performance irrespective of the signal wavelength chosen. As noted above, the search for new materials to resolve this conflict continues in many research laboratories.

11.7 FIBRE AND CABLE MANUFACTURE

The objective of fibre manufacture is to fabricate a single fibre with a high refractive index core and a lower refractive index cladding. The first step is to make a *preform*. This is a rod of diameter of 15–100 mm and a length of 1–1.5 m and a refractive index profile, from centre to outside, identical to that needed in the finished fibre. The refractive index profile within the preform is achieved by creating one or more layers of doped silica glass on the inside of the tube. In particular, replacement of Si by the heavier Ge is used to increase the refractive index. This poses no great problems because chemically GeO_2 is very similar to SiO_2 and the two oxides readily blend together to form a homogeneous solid solution.

The starting point in the manufacture of a preform is a high purity silica tube containing only a few parts per million of hydroxyl groups. This is rotated and heated while a gas consisting of various amounts of $SiCl_4$, $GeCl_4$ $POCl_3$, Freon and oxygen is allowed to flow through its centre as in Figure 11.9. At the temperatures of the tube, 1700–1900°C, the gases decompose thus:

$$SiCl_4 + O_2 \rightarrow SiO_2 + 2Cl_2$$
$$GeCl_4 + O_2 \rightarrow GeO_2 + 2Cl_2$$
$$4POCl_3 + 3O_2 \rightarrow 2P_2O_5 + 6Cl_2$$

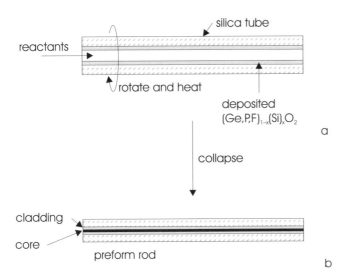

Figure 11.9 The formation of a preform rod for silica fibre production. (a). Reactive gases are passed through the centre of a hot rotating silica tube, where they decompose to form layers of silicon dioxide glass, $(Ge,P,F)_xSi_{1-x}O_2$, containing varying amounts of dopant, so as to give the cladding (lower dopant content) and core (higher dopant content). After reaction, increased heating causes the tube to collapse to the preform (b)

The result is that a "soot" consisting of a mixture of silicon, germanium and phosphorus oxides forms on the inside of the tube. As the heating zone traverses the tube the "soot" merges with the tube to form a glass inner coating about 10 μm thick. The ratio of the dopants can be changed by changing the ratio of the chlorides in the initial mixture and this allows the refractive index of the layer to be varied in a controlled manner.

The refractive index variation is achieved in two stages. Of the order of 12 to 32 layers are deposited initially to form an inner coating inside the tube which will become the cladding on the fibre. This has a composition of SiO_2 containing 0.1–1% P, F and Ge, the fluorine incorporation arising from the Freon gas present. The refractive index is close to that of pure SiO_2. Following this stage the material which will ultimately form the core is deposited. This is achieved by depositing 4 to 10 layers of material with an overall composition somewhere between the limits $(Ge,P,F)_{0.06}Si_{0.94}O_2$ to $(Ge,P,F)_{0.3}Si_{0.7}O_2$, depending upon final use. As a result of the higher germanium content, this deposit has a higher refractive index than the first.

When sufficient layers have been formed the temperature is raised enough to cause the tube to collapse under surface tension. The result is a solid rod with a centre of higher refractive index glass surrounded by a region of lower refractive index glass enveloped in the original silica of the tube. This solid rod is the preform.

To transform the preform into a fibre, a process called *fibre drawing* takes place, in which the end of a preform rod is softened to near to its melting point. Under these conditions glass has the property that it can be pulled out and will form a long fibre. Surprisingly the refractive index profile of the preform is preserved exactly in the fibre even though the preform diameter of 15–100 mm is drawn down to approximately 1/10 mm. The procedure is schematically depicted in Figure 11.10.

As the fibre is withdrawn the preform is slowly lowered into the furnace so as to maintain continuous production. The technology of the process is important. Cleanliness of the environment is vital as any particles deposited on the fibre surface reduce its strength. It is also vital that the furnace has a stable temperature, otherwise the fibre dimensions will vary. Similarly, great precision is needed in winding the fibre onto a drum in order to avoid breaks. Moreover, at all stages the dimensions and concentricity of the fibre must be monitored. This process is assisted by the unusual feature that silica glass melts to a Newtonian fluid. In this case the diameter decrease is proportional to the force applied, not the stress, which is equivalent to force per unit area.

Freshly drawn fibres have a strength similar to that of steel. However, this is rapidly degraded by surface reaction, especially with water vapour. To offset this the fibre must be coated with a thin polymer film. The material chosen is frequently a silicone resin. This has the advantages of providing mechanical strength, whilst curing rapidly and producing a low "tack" finish so that fibre can be wound on and off the final spool.

These coatings are called *primary coatings*. *Secondary coatings* which are applied for added protection are often colour coded and are usually added later. The commonest materials used for this purpose are urethane acrylates or nylon.

The final step in the manufacturing process is to cable the individual fibres. The primary reason for this is to improve mechanical strength. A second reason is that a multiplicity of fibres is required for systems and it is easier to bundle them together in one cable. The cable design depends upon the ultimate use for the link and its geographical location. Cables for battlefield or rugged terrain use must have different mechanical properties to those used in office buildings or in underwater links.

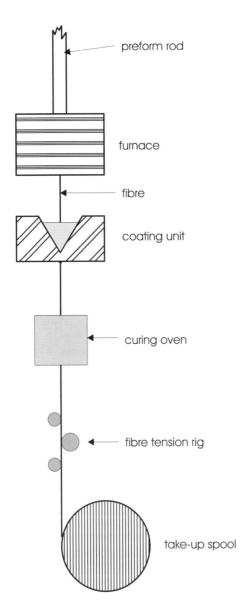

Figure 11.10 The major components of a fibre drawing "tower". The preform is softened in a carefully designed furnace and drawn into a fibre. It is immediately covered by a polymer in the coating unit which is then hardened in the curing oven before being wound onto a spool

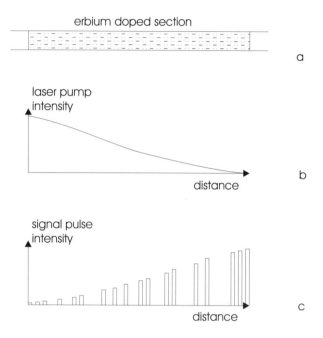

Figure 11.11 Schematic illustration of the transfer of energy from a laser pump to signal pulses as they traverse a length of erbium doped optical fibre. (a) A silica fibre containing a length doped with Er^{3+} ions. (b) The intensity of the laser pump radiation falls as it traverses the doped region as energy is lost by excitation of the Er^{3+} ions. (c) The signal pulses gain energy from the excited Er^{3+} ions as they traverse the doped region

11.8 OPTICAL AMPLIFICATION

The amplification of signals in fibre-optic transmission systems is of great importance. Originally amplification used costly *repeaters*, which transformed the optical pulses into electronic signals, amplified these and then recreated optical pulses. Operating systems are now available which use a section of optical fibre doped with erbium (Er^{3+}) as the activator. The amplifying section consists of about 30 m of monomode fibre core containing just a few hundred parts per million of Er^{3+}. This section of the fibre is illuminated by a semiconductor diode laser (see Chapter 13) so as to excite the Er^{3+} ions. The commonest wavelengths used are 980 nm and 1480 nm. The erbium ions have the remarkable ability to transfer energy from the laser to the signal pulses as they traverse this section of fibre. The result is shown schematically in Figure 11.11.

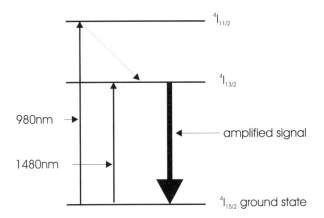

Figure 11.12 Schematic diagram of the important energy levels and transitions in erbium optical amplifiers. The transition produced by a pump frequency of 980 nm rapidly loses energy and drops to the $^{4}I_{13/2}$ level, which is also filled by a pump frequency of 1480 nm. (The energy levels shown as single lines consist of groups of closely spaced levels.) The energy of the $^{4}I_{13/2}$ level is transferred to the signal pulses (heavy arrow) as they traverse the doped region of fibre

The energy transfer comes about in the following way. With reference to the schematic energy level diagram of the Er^{3+} ion is shown in Figure 11.12 we see that illumination of the erbium containing section of fibre with energy of wavelength 980 nm excites the ions from the ground state ($^{4}I_{15/2}$) to the upper state ($^{4}I_{11/2}$) from whence they rapidly decay to the $^{4}I_{13/2}$ level. This process is referred to as *pumping* and the laser involved as the *pump*. The use of radiation of 1480 nm wavelength excites the Er^{3+} ions directly from the ground state to the $^{4}I_{13/2}$ level. This state has quite a long lifetime. A passing light pulse, with a wavelength close to 1480 nm, empties the Er^{3+} excited state via stimulated emission. (See Chapter 13 for more information on this process.) In effect, as we said, the pump energy is transferred to the signal pulses over the course of the erbium doped stretch of fibre. This achieves signal amplification while retaining the coherence of the pulse constituting the signal. The amplification is impressive and a 10 mW pump laser can give a thousandfold increase in signal power.

11.9 SECOND HARMONIC GENERATION AND COLOUR PRODUCTION IN SILICA FIBRES

Imagine the following situation. Intense infrared laser light pulses with a wavelength of 1060 nm from a Nd^{3+}:YAG laser (see Chapter 13) are being sent down an ordinary monomode fibre of the type described above. After an hour or so a green light with a wavelength of 530 nm starts to appear along with the input infrared. As time goes by the intensity of the green light increases and after 10 h or so is quite prominent. It looks as if we have frequency doubling and efficient second harmonic generation taking place.

As discussed in Chapter 3 a non-centrosymmetric (uniaxial or biaxial) crystal is needed for this effect to be observed. Second harmonic generation is ultimately dependent upon a non-zero second-order non-linear dielectric susceptibility coefficient $\chi^{(2)}$ in the polarisation equation:

$$P = \varepsilon_0 \chi^{(1)} \, E_0 \cos \omega t + \varepsilon_0 \chi^{(2)} \, (E_0 \cos \omega t)^2 + \varepsilon_0 \chi^{(3)} \, (E_0 \cos \omega t)^3 + \ldots$$

Glasses are isotropic and the value of $\chi^{(2)}$ for any glass is always is zero. A glass, therefore, should not give rise to second harmonic generation. How is it that a non-zero second-order dielectric susceptibility term has evolved during the infrared irradiation?

The answer to the puzzle required a great deal of careful experimental work and not all of the processes taking place at an atomic level have yet been clarified. However, the major sequence of steps occurring seems to be these. The intense electric field of the infrared radiation is strong enough to cause ions in the glass to migrate in a process akin to ionic conductivity. In the region near to the core–cladding interface the ionic displacements are different in the cladding than in the core. This results in permanent charge separation. The resulting permanent electric field that builds up in the boundary region, E_{dc}, has been found to be as large as 10^8 V m^{-1}. It is this intense field that is responsible for the second harmonic generation. Interestingly the value of $\chi^{(2)}$ is still zero, but an "effective" $\chi^{(2)}$ term, given by the product $E_{dc}\chi^{(3)}$ is present at the core–cladding interface.

The actual species which migrate are not yet clarified, but Ge^{4+}, P^{5+} and B^{3+} dopant ions in the core region, and impurities such as OH^-, H^+ and Na^+ may all have a part to play. This is supported by the observation that silica glass plates subjected to about 3 kV and heated to a temperature of 250–325°C also show the same second harmonic generation effect in a region of about 3 μm close to the surface adjacent to the positive contact. This region has been found to be depleted in ionic constituents and it seems that ionic movement in the applied voltage establishes a permanent electric field at the surface. Heating the glass at a temperature of a few hundred degrees in the absence of an applied voltage allows for ionic diffusion to re-occur, cancels the effect and returns the glass to its original non-active state.

11.10   ANSWERS TO INTRODUCTORY QUESTIONS

Why was it necessary to change to optical communication links?
The primary reason was that existing copper cable links were full. This refers both to the carrying capacity of each copper cable and also the ducting carrying the cables. Of the alternatives, optical fibres seemed to be the best.

What technology was needed to set up optical links?
The basic enabling technology needed was at least (i) pure glass, (ii) a high power reliable monochromatic light source, (iii) optical encoders and decoders, (iv) connectors and couplers and (v) amplifiers or repeaters.

What are the chemical and physical restrictions on optical links?
The physical restrictions are infrared absorption at lower energies, electronic absorption at higher energies and Rayleigh scattering. These impose a limit on the attenuation versus wavelength curve for the material used. In silica fibres these physical limits have been reached. Chemically impurities of any type degrade performance by spurious absorption. In silica fibres the major problem remains hydroxyl, $-OH$.

What future developments in optical links might be anticipated?
Now that "in fibre" optical amplifiers are available (a recent development), the major improvements will be new materials. Heavy metal fluoride glasses show promise because they should have an attenuation of about one hundredth that of silica, but they are chemically very difficult to make.

How can doped fibres be made to amplify optical signals?
Doped fibres which act as amplifiers consist of a length of silica fibre containing Er^{3+}. These ions are excited to higher energy levels using a suitable diode laser. The lifetime of the excited state is such that they are depopulated by stimulated emission (see Chapter 13) by each passing signal pulse. Both the output pulse and the stimulated signal are coherent and of the same wavelength and the signal is thus amplified.

11.11   FURTHER READING

An account of the optics of fibre optic communications is to be found in:

O. S. Heavens and R. W. Ditchburn, Insight into Optics, Wiley, New York (1991), Ch. 14.

The evolution of fibre optic communications can be appreciated by reading the series of articles:

W. S. Boyle, Scientific American, **237**, August 1977, p. 40.
A. Yariv, Scientific American, **240**, January 1979, p. 54.
M. G. Drexhage and C. T. Moynihan, Scientific American, **259**, November 1988, p. 76.
E. Desurvire, Scientific American, **266**, January 1992, p. 96.

Information on erbium optical amplifiers will be found in:

P. G. Kik and A. Polman, MRS Bulletin, **23**, April 1998, p. 48 and references therein.

The frequency doubling effect in silica fibres and plates is described by:

W. Margulis, F. C. Garcia, E. N. Hering, L. C. Guedes Valente, B. Lesche, F. Laurell and I. C. S. Carvalho, MRS Bulletin, **23**, November 1998, p. 31 and references therein.

11.12 PROBLEMS AND EXERCISES

1. Estimate the attenuation of a fibre in which 50% of the input power is lost over a distance of 1 km.

2. Window glass has an attenuation of about 10^6 dB km^{-1}. What thickness of this glass would block a laser beam of 10^9 W cm^{-2}?

3. Light from a laser with a wavelength 800 ± 20 nm is launched into a silica fibre of length 50 km. What will be the spread of the exit pulse (in metres) if the refractive index of the silica is 1.45363 at $\lambda = 780$ nm and 1.45298 at $\lambda = 820$ nm? What is the dispersion?

4. A silica fibre uses a 1 mW laser operated at 10^8 pulses s^{-1}, has an attenuation of 0.25 dB km^{-1} and a detector that can register pulses of 10^{-16} W. What is the maximum length of fibre before amplification is needed?

5. Estimate the modal dispersion of an instantaneous light pulse transmitted in a stepped index fibre with a core refractive index of 1.51 and a cladding refractive index of 1.49.

CHAPTER 12

DISPLAYS

How do TV screens produce colour?
What are LEDs?
How do electroluminescent displays produce
coloured images?
By what mechanism do active matrix liquid crystal
displays operate?
How do plasma displays generate colours?

Undoubtedly the display device that has dominated the second half of the 20th century is the television or cathode-ray tube (CRT). This is still the most widespread display system for information and entertainment at the end of the century. However, a major shortcoming with CRT technology is that the screen size is limited to a diagonal of about 900 mm. Although not a problem for present day analogue transmissions this limit will be noticeable in the case of high definition digital television. In addition the tubes are bulky and rather heavy.

The newer device technologies described below are all, in one way or another, successfully replacing CRTs for particular applications. These first gained prominence in portable devices such as calculators and watches, with the initially used light emitting diodes (LEDs) being replaced in turn by liquid crystal displays (LCDs). The aim of this chapter is to outline the way in which these and other display devices turn electrical input into visible, usually coloured, patterns of information.

12.1 CATHODOLUMINESCENCE AND CRT DISPLAYS

Television tubes, also called CRTs[7], function by using electron impact to excite a phosphor. The electrons are produced by heating a metal filament, which forms the cathode of the device, in an evacuated glass envelope.

---

[7] Experiments at the end of the last century concerned with the action of electric fields on gases at low pressures in glass tubes led to the discovery of a new sort of radiation, *cathode rays*, which were emitted by the cathode. Although Thomson showed cathode rays to be electrons in 1897, the name cathode ray tube is still normally used in the literature.

Figure 12.1   The arrangement of red (r), green (g) and blue (b) phosphor dots on a colour cathode ray TV tube

They are accelerated towards a cylindrical anode by the application of a high voltage. The anode and cathode assembly is often referred to as an *electron gun*. The electrons emerge from the anode as a narrow collimated beam which is moved horizontally and vertically by electrostatic or electromagnetic means. A vacuum is needed because electrons cannot pass through a gas without frequent collisions and subsequent loss of energy. Phosphors are deposited internally on the front face of the tube to form the screen. The impact of the highly accelerated electrons on the phosphors produces visible light. This process is known as *cathodoluminescence.*

The electron beam is scanned across and down the screen in a pre-determined pattern. As the focused beam sweeps past a dot of phosphor, light is emitted. The screen is therefore lit up by small spots of phosphor whose emission is refreshed at each pass of the electron beam. In both monochrome and colour television the light emitted is perceived by the eye in the same way that impressionist pointillist paintings are, by way of additive coloration. For colour television three primary colours are used, arranged in an array as depicted in Figure 12.1. Just as in pointillist paint-ings, it is important that the phosphor spots do not overlap in colour displays, otherwise the picture quality is degraded.

This brief summary conceals a vast amount of research on CRT displays. In practical terms the *efficiency* of the phosphor is of maximum import-ance. The efficiency of a fluorescent material depends upon many factors including whether an activator alone is needed or whether a sensitisor and an activator are involved. The concentrations of the activator and sensi-tisor play an important part in controlling efficiency and optimal con-centrations have to be found experimentally. Impurities generally have a negative effect and high purity is a necessity.

A further requirement is a knowledge of the *decay time* of the phosphor. The intensity of the light given out by a phosphor after excitation is removed is frequently given by an equation of the type:

$$I_t = I_0 \exp(-t / \tau)$$

where I_t is the intensity after time t has elapsed, I_0 is the initial intensity the moment excitation ceases and τ is the decay time. The decay time is the time taken for the luminescent radiation to decay to $1/e$ of its initial value. This depends on a variety of factors including the quantum mechanical selection rules which control the energy transitions in the solid. Clearly, for a continuous picture to be observed, the decay time must be longer than the time that it takes the electron spot to complete its round trip and refresh the emission again. However, this must not be so long that shadow images persist after the action has moved on. In this respect *after glow* can be troublesome. After glow is said to occur when the luminescence decays at the rate expected until a certain value is reached and then decays much more slowly. This is frequently due to the presence of impurities in the material which contribute differing mechanisms to the emission process.

In summary, a phosphor must have: a high efficiency; a suitable decay time; a suitable emission spectrum; and a low after glow level.

Black-and-white TV uses silver-activated zinc sulphide (Ag:ZnS) to give a blue colour plus silver-activated zinc-cadmium sulphide (Ag:(Zn,Cd)S) to give a yellow. The ratio of these two phosphors controls the overall hue of the screen. Colour TV also uses silver-activated zinc sulphide (Ag:ZnS) for the blue colour, copper-activated zinc-cadmium sulphide (Cu:(Zn,Cd)S) for green and europium-activated yttrium oxysulphide ($Eu^{3+}:Y_2O_2S$) for red. The phosphor layers are backed by an aluminium film to increase brightness of the image.

There are a number of other devices which use CRT technology widely. Most familiar are computer monitors, flying spot scanners, oscilloscopes and radar screens. These need different decay characteristics than TV screens and the phosphor technology in each is tailored to the exact requirements of the product.

12.2 SEMICONDUCTOR LIGHT EMITTING DIODES

Semiconductor light emitting diodes (LEDs) were first widely used in displays on calculators and similar electronic instruments. Recently they have become widespread as rear brake-light strips on cars, where the red LED colour contrasts with the better known brake-light red of the coloured plastic lenses used traditionally.

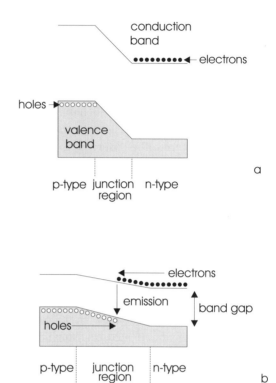

Figure 12.2 (a) The equilibrium situation in a p-n junction. Under this condition a charge builds up in the junction region which prevents holes and electrons recombining. (b) Under forward bias, when a positive voltage is applied to the p-type region, electrons and holes enter and recombine in the junction region, leading to emission of radiation with an energy approximately equal to the band gap. For simplification, the acceptor levels and donor levels which are located in the band gap are not shown

Semiconductor light emitting diodes are formed by the juxtaposition of a region of n-type and p-type semiconductor, grown into a single crystal, as sketched in Figure 12.2a. In the p-type region of the material the semiconductor has been doped with acceptors so that the top of the valence band contains a high population of mobile holes. In the n-type region the semiconductor has been doped with donors, so that the material contains a population of mobile electrons at the bottom of the conduction band. Under normal conditions an equilibrium is reached which keeps the holes and electrons confined in their own halves of the junction.

When a positive voltage (called a *forward bias*) is applied to the p-type side of the junction, the equilibrium barrier height falls and the junction

Plate 9.3 (a) A pale yellow phosphor based on zinc sulphide viewed in normal daylight

(b) The same material irradiated with ultraviolet light with a wavelength range of from 350nm–380nm, showing a bright green-yellow fluorescence

Plate 10.1 Crystals of the mineral vermilion (mercuric sulphide, HgS)

(a)

(b)

Plate 10.2 Crystals of the mineral chalcopyrite (copper iron sulphide, $Cu_2Fe_2S_4$), containing (a) smaller and (b) greater amounts of copper

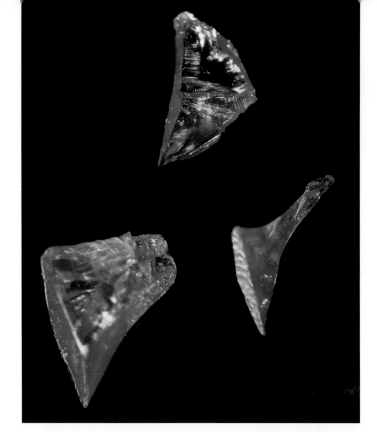

Plate 10.3 Crystals of rutile, one form of titanium dioxide, TiO_2, coloured by small quantities of gallium trioxide, Ga_2O_3, impurity, which cause hole colour centres to form in the rutile matrix

Plate 12.1 The liquid crystal display in a calculator dating from ca. 1975. The dark colour is due to the blocking of polarised light by a thin liquid crystal layer sandwiched between transparent electrodes

region broadens, as indicated in Figure 12.2b. Electrons and holes now enter the junction and recombine. The energy released, which is approximately equal to the band gap, appears as light.

There are a number of aspects of importance for a working device. Clearly the band gap must be such as to give out visible wavelengths. This can be achieved using solid solutions to manipulate the band gap, as described in Chapter 10. The most widely used LED materials are solid solutions between gallium arsenide (GaAs), gallium phosphide (GaP), aluminium arsenide (AlAs) and aluminium phosphide (AlP). By varying the amounts of each of these elements in the solid solution, LEDs which emit light from the red to yellow can be produced.

A second consideration is important. Ideally LEDs must be highly efficient and give an adequate light output under a small voltage. It is for this reason that GaAs based materials are preferred. The transition from the valence band to the conduction band in this compound is a *direct transition*, as opposed to that in silicon, which is an *indirect transition*. *Indirect transitions* can only take place if some energy is simultaneously lost to lattice vibrations. Materials showing direct band gap transitions are far more efficient than those utilising indirect band gap transitions.

The reason why gallium arsenide solid solutions are limited to the red/yellow colours is due to the fact that the transitions become indirect when the band gap becomes appropriate to green or blue emission. To overcome this problem gallium nitride (GaN) and related materials are being explored as blue/violet emitting LEDs especially with the aim of producing a blue diode laser, as discussed in Chapter 13.

Full colour displays using LEDs have been built using small red, green and blue LEDs in each of thousands of pixels. The eye perceives additive coloration as with a TV screen. These displays have no size limitations and the intensity produced by each LED is also sufficient to make them easily visible in daylight. The major limitation remains that of cost. Although each individual LED is inexpensive, the large numbers of LEDs incorporated into a high definition display result in an expensive unit.

12.3 PHOSPHOR ELECTROLUMINESCENT DISPLAYS

Electroluminescence is the general term which is used to refer to the emission of light from a recombination of electrons and holes in a suitable material following the application of an appropriate voltage. Thus direct

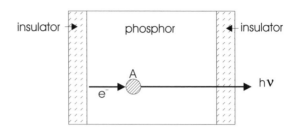

Figure 12.3 Diagrammatic representation of the process taking place in an electro-luminescent material. Electrons enter the phosphor from an insulator/phosphor interface and accelerate under a high voltage. These transfer energy to luminescent centres, A, via collisions and these in turn lose energy by emitting light

bandgap semiconductors such as gallium arsenide are electroluminescent materials. Newer developments exploit phosphors and organic molecules or polymers as the active material in a variety of light emitting devices.

Electroluminescent displays containing a thin film of a phosphor are rather similar to CRTs except that the excitation of the phosphor is by an applied voltage rather than by direct electron impact. The principle behind the operation of a phosphor electroluminescent display is illustrated in Figure 12.3. Electrons enter the phosphor at the junction with a surface insulating coating and are accelerated under the influence of a high electric field. These electrons collide with the luminescent centres in the phosphor, transferring energy in the process. The excited luminescent centres then fall back to the ground state and release energy by light emission.

These devices are rugged and thus find favour for military use. Mono-chrome displays are now widely used as display panels in a number of military projects. Phosphor electroluminescent panels are also used as backlighting in small units such as digital watches.

The most promising devices use *a.c.* supplies in a *thin film electro-luminescent* (ACTFEL) display. A monochrome display has a structure similar to that shown schematically in Figure 12.4. The device is built up in thin layers on a glass substrate. A layer of a transparent electrical con-ductor, most often indium tin oxide, of about 400 nm thickness, is laid down first. This material acts as one electrode. This is then covered with about 400 nm of a transparent insulator. The active layer, about 700 nm of phosphor and another 400 nm layer of transparent insulator are then added. Finally, a 200 nm thick layer of aluminium is deposited on the stack. This serves as an electrode and also reflector. The display is viewed through the glass substrate, which acts as a protective surface.

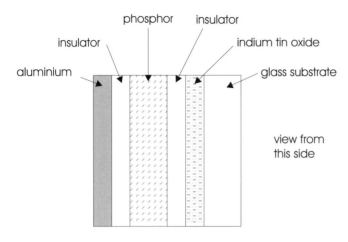

Figure 12.4 The arrangement of materials in one pixel of a monochrome ACTFEL (acthin film) electroluminescent display

The arrangement is, in fact, a series of capacitors. This design is chosen because the high voltages which are needed in the phosphor layer are generated from low electrode voltages by way of the capacitance of the thin insulating layers.

The most efficient electroluminescent thin film phosphors consist of zinc sulphide containing manganese (Mn^{2+}:ZnS) as the luminescent centre. These emit yellow 585 nm light from d–d transitions within the Mn^{2+} ion due to the crystal field splitting of the d-orbitals, as explained in Chapter 7. A number of other colours can be generated using the f–f transitions of lanthanide ions. Red is produced by calcium sulphide doped with europium (Eu^{2+}:CaS), green by zinc sulphide doped with terbium (Tb^{2+}:ZnS) and blue with zinc sulphide doped with both thulium and fluorine (Tm^{3+},F^-:ZnS). In this latter case the thulium ions (Tm^{3+}) occupy positions normally filled by zinc (Zn^{2+}) ions. The fluoride (F^-) ions are needed to compensate for the excess charge on the thulium ions; ($Tm^{3+} + F^-$) being equivalent to ($2Zn^{2+}$). Broad band white light emission can also be achieved using combinations of luminescent centres in a single host matrix.

Full colour displays can be constructed in two ways. One way is to use a white light emitting phosphor and to incorporate a series of colour filters into the device. A more interesting way is to build up subarrays of pixels with red, blue and green phosphor subpixel units, as sketched in Figure 12.5. By varying the voltage distribution between the aluminium and

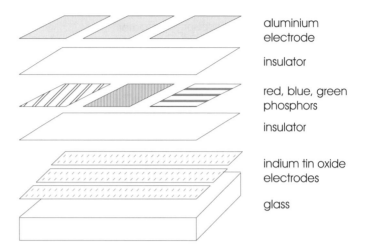

aluminium
electrode

insulator

red, blue, green
phosphors

insulator

indium tin oxide
electrodes

glass

Figure 12.5   Diagrammatic arrangement of the layers in one pixel (three subpixels) of
a full colour ACTFEL electroluminescent display

indium tin oxide electrodes any of the red, blue or green phosphors can be
excited to luminesce.

12.4   ORGANIC ELECTROLUMINESCENT DISPLAYS

Organic dye molecules, either alone or in polymeric forms, can also be used
in electroluminescent displays. The important energy levels are the $\pi - \pi^*$
pair shown in Figure 8.1 and noted when we discussed colour production by
organic molecules. Although the upper energy level is normally empty, by
using suitable electrodes the upper (the π^* level) and lower (the π level)
energy states can be continually filled by electrons and holes. These can
then recombine to produce the colour associated with the dye molecule.

Most organic dyes will not conduct either holes or electrons although
they can usually be persuaded to conduct one of them by suitable doping.
The resultant materials are known as p-type if they conduct holes and n-
type if they conduct electrons, by analogy with inorganic semiconductors.
Devices are then constructed of layers of p- and n-type materials in contact,
as depicted in Figure 12.6. These are sandwiched between a negative
(electron emitting) electrode, most often an alloy of silver and magnesium,
and a conducting but transparent positive (hole emitting) electrode, usually
indium tin oxide. The light emitting region is close to the junction of the two

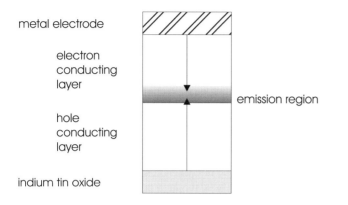

metal electrode

electron
conducting
layer

emission region

hole
conducting
layer

indium tin oxide

a

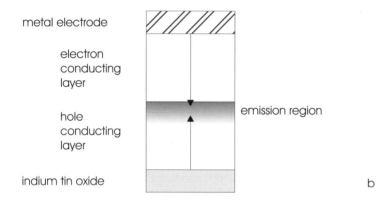

metal electrode

electron
conducting
layer

emission region

hole
conducting
layer

indium tin oxide

b

Figure 12.6 Typical structure of double layer molecular thin film electroluminescent devices. In (a) the light emitting region is in the electron conducting layer and in (b) it is in the hole conducting layer

dye layers. As with many electronic devices, cleanliness during manufacture is vital and the best devices are those in which the dye molecules are laid down by evaporation and condensation on the device surface carried out in a high vacuum.

Conjugated polymers can also be used in these devices and a number of polymer LEDs have been produced. The flexibility of the polymerisation process makes it possible to produce copolymers of two units, one of which is inherently electron conducting and one of which is hole conducting. Thus a typical polymer LED might consist of a layer of conjugated polymer sandwiched between two electrodes and deposited on a glass substrate. The energy band diagram for an electroluminescent display using a polymer of the organic molecule (2-methoxy-5-2'-ethyl-hexyloxy)-1,4-polyphenyl vinylene[8], Figure 12.7a, more conveniently referred to as MEH-PPV, is shown in Figure 12.7b. It has been found that suitable electrode materials are indium tin oxide and calcium metal. At the left hand side of the diagram the energy level represents the energy needed to move holes into the polymer π level from indium tin oxide electrode and at the right the energy level shows the energy needed to move electrons from the calcium electrode into the polymer π^* level. These polymeric layers have the advantage that the colour emission can be tuned easily by the tailoring of sidegroups on the polymer backbone.

Because of the vast number of organic molecules available the range of colours that can be obtained is enormous and full colour displays can be produced. An additional advantage is that these materials are light in weight and flexible. Thus it is possible to imagine large area roll-up displays which can be easily transported from one location to another and simply hung on a wall. The display type is also well adapted for large area back-lights. Although there are no commercially available devices at present they are likely to come into production soon.

12.5 LIQUID CRYSTAL DISPLAYS

Liquid crystal displays (LCDs) are commonplace on portable calculators, digital watches and some laptop computers as black-on-grey images. A typical display character is shown in Plate 12.1. Use of colour filters has extended this technology to include colour displays. Portable computers and digital cameras now use such displays routinely. However, there are problems in extending the technology to large displays and current interest is in the production of large area flat panel displays using active matrix liquid crystal display (LCD) technology.

As we saw in Chapter 3, liquid crystals are not, themselves, coloured. However, the molecules present in the liquid crystal have two important and exploitable properties. The molecules can alter the orientation of the

---

[8] The outmoded term term phenylene (bivalent C_6H_4) is still used in the literature instead of the more correct term phenyl, leading to a nomenclature of (2-methoxy-5-2'-ethyl-hexyloxy)-1,4-phenylene vinylene.

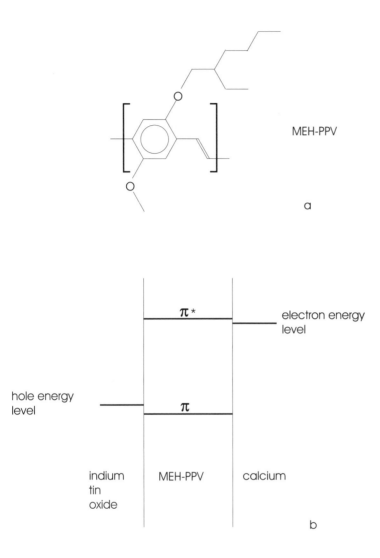

MEH-PPV

a

π *

electron energy
level

hole energy
level

π

indium
tin
oxide

MEH-PPV

calcium

b

Figure 12.7 (a) The skeletal representation of the structure of the polymeric material
MEH-PPV. The unit in brackets is repeated many times in a polymer molecule. Each
line segment represents a C-C bond and all atoms are omitted except for oxygen (O).
(b) The energy level structure for polymeric MEH-PPV, coupled with calcium and
indium tin oxide electrodes

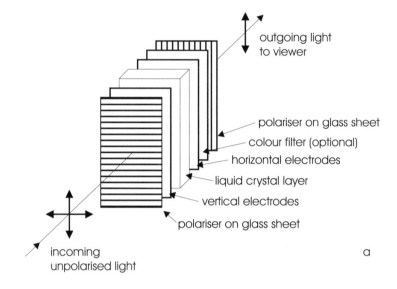

outgoing light
to viewer

polariser on glass sheet
colour filter (optional)
horizontal electrodes
liquid crystal layer
vertical electrodes
polariser on glass sheet

incoming
unpolarised light

a

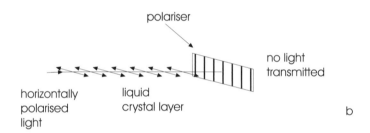

polariser

no light
transmitted

horizontally
polarised
light

liquid
crystal layer

b

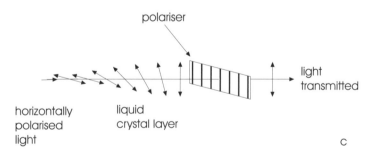

polariser

light
transmitted

horizontally
polarised
light

liquid
crystal layer

c

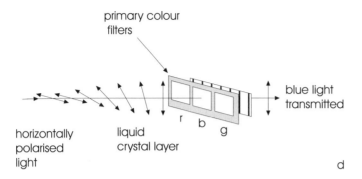

Figure 12.8 (a) The diagrammatic construction of a liquid crystal display. Unpolarised light is produced by fluorescent lighting. When a pixel is turned off (b) the liquid crystal film does not rotate the plane of polarisation of the light and it does not emerge from the display. When the pixel is turned on (c) the plane of polarisation is rotated and light is able to pass the second polariser and emerge. (d) Three colour filters can be interposed to make up three subpixels in order to create a full colour display

direction of polarisation of light and the alignment of the molecules is easily changed by an externally applied electric field. These attributes apply to the liquid crystal film which is at the heart of the display. The film is sandwiched between two glass sheets which are bounded by crossed polars, as drawn in Figure 12.8a. White unpolarised background lighting provides the illumination. One of the glass sheets is overlaid with a set of vertical electrodes while the other with a horizontal set. The application of a voltage across any pair of vertical and horizontal electrodes allows any pixel in the display to be activated. When no voltage is applied to the electrodes the polarised light beam is not rotated on passing through the liquid crystal layer, as in Figure 12.8b. In this case the light is blocked by the second polariser and the pixel appears dark. Applying a voltage to the electrodes causes the liquid crystal molecules to rotate and in so doing to change the plane of polarisation of the linearly polarised light beam by 90° as shown in Figure 12.8c. The polarised light "follows" the molecules and the plane of polarisation is rotated. It is then passed by the second polar and so the pixel looks bright.

Colour displays operate in a similar way. Each pixel is composed of three *subpixels*, each of which has a colour filter imposed before the final polariser. The fabrication of these colour filters is intricate and involves the dispersion of an organic pigment in a clear polymer substrate. In order to obtain a full colour image three different pigments need to be used, corresponding to the three primary colours. Each of these is allocated to its appropriate subpixel. In devices, each subpixel is controlled by the electrode grid, allowing it to be on or off independently of others in the array. The colour seen is thus additive and the primary colours used are red, green and blue. The display is usually viewed against a black background which enhances the colour contrast.

In the case of active matrix liquid crystal displays the grid of ruled electrodes is replaced with a grid of transistors. One transistor controls one pixel. While these displays are of high quality, the expense of creating the transistor array is high and large active-matrix display panels are not available commercially. Larger displays are presently reserved for applications where cost is of secondary importance, as in some military applications.

12.6 PLASMA DISPLAYS

Flat-screen plasma displays in monochromatic form were first used in some portable computers. These used an ionised gas to produce an orange-red colour in a similar fashion to that exploited by neon lights. The gas was confined in a series of wells and two grids of transparent electrodes, one running horizontally and one vertically, provided the necessary current and voltage to ionise the gas. A monochrome display of this type is shown in Plate 7.1. Colour plasma flat panel displays are similar but use phosphors to produce light rather than ionised gas, thus recalling CRT displays. The technology, however, is quite different to that in the CRT and has much more in common with the fluorescent light technology described in Chapter 9.

The general structure of a colour plasma display is depicted in Figure 12.9a. It consists of a pair of glass plates containing a series of wells bounded at the top surface and the bottom by electrodes. In order to form a display the region between the glass sheets is divided up into pixels by a series of ribs or separators controlled by two sets of electrodes arranged at right angles to each other. Each different phosphor cell forms a subpixel, three of which combine to make a full colour pixel. Each well is coated internally with a red, green or blue phosphor. The structure of a well is shown in more detail in Figure 12.9b. The layer of magnesium oxide (MgO) serves as a dielectric to enhance the electric field present in the well.

The working gas in the wells is a mixture of neon and xenon. When a high voltage is applied across the two electrodes above and below a well the gas is excited and releases the energy gained as ultraviolet radiation. The ultraviolet light excites the phosphors to emit in the red, green or blue, in the same way that the ultraviolet light from a mercury vapour excites

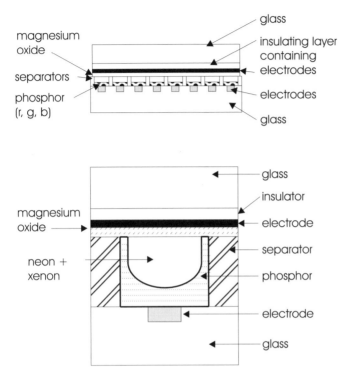

Figure 12.9 (a) The layer structure of a full colour plasma display. Each well in the structure contains a red, green or blue phosphor. The arrangements of the upper and lower electrodes perpendicular to one another creates an array of subpixels. (b) Detail of one subpixel. A mixture of neon and xenon emits ultraviolet radiation which interacts with the phosphor to give out either red, green or blue light

the phosphors in a fluorescent tube. The main phosphors used at present are a mixed rare earth borate doped with europium (Eu^{3+}:(Y,Gd)BO_3) which gives a red emission, a complex aluminate doped with europium (Eu^{2+}:BaMgAl$_{14}O_{23}$) for blue emission and zinc silicate doped with manganese (Mn^{2+}:ZnSiO$_3$) for green light.

12.7 ANSWERS TO INTRODUCTORY QUESTIONS

How do TV screens produce colour?
TV screens produce colour by electron impact on a phosphor. The energy from the electron impact is converted to light output using fluorescent centres.

What are LEDs?

LEDs are light emitting diodes. They are frequently made from a slice of gallium arsenide single crystal which has been doped to produce an n-type conducting region adjacent to a p-type conducting region. Light is emitted in the junction between these two regions when a forward bias is applied to the device. The light is the energy released by the recombination of electrons and holes in the junction.

How do electroluminescent displays produce coloured images?

Electroluminescent displays can use either phosphors or organic molecules to produce light. When phosphors are used, energetic electrons in the phosphor collide with luminescent centres in the phosphor which converts the energy to light. Phosphors can produce coloured images by using arrays of red, blue and green emitting materials. In organic displays, electrons and holes occupying $\pi - \pi^*$ orbitals recombine to release light. To form a colour display, different dye molecules are chosen to form pixels with different colour emission.

By what mechanism do active matrix liquid crystal displays operate?

Active matrix liquid crystal displays use a liquid crystal film to rotate the plane of linearly polarised light so as to allow it to pass through a second polar sheet. The ability of the liquid crystal to rotate the light varies with an applied voltage, so that the intensity of light emerging from the pixel varies as a function of the applied voltage.

How do plasma displays generate colours?

Colour plasma displays use ultraviolet radiation generated by electron bombardment of a He / Ne gas mixture to excite a phosphor. The energy is released in the form of light. Each pixel contains three subpixels which emit three primary colours, red, green and blue, which combine to give a full colour pixel.

12.8 Further Reading

A survey of competing flat panel display types with emphasis on plasma displays is given in:

A. Sobel, Scientific American, **278**, May 1998, p. 48.

A simple description of active matrix liquid crystal displays will be found in:

S. Musa, Scientific American, **277**, November 1997, p. 87.

More technical information is given in:

MRS Bulletin, **23**, March 1996, which includes:
Y. Yang (Polymer Electroluminescent Devices), p. 31.
J. Hanna and I. Shimizu (Active Matrix Liquid Crystal Displays), p. 35.
T. Tsutsui (Molecular Thin Films), p. 39.
P. D. Rack, A. Naman, P. H. Holloway, S.-S. Sun and R. T. Tuenge (Electro-luminescent Displays), p. 49.
L. F. Weber and J. D. Birk (Colour Plasma Displays), p. 65.

The luminescent phosphors used in CRTs and other devices are described by:

G. Blasse and B. C. Grabmaier, Luminescent Materials, Springer-Verlag, Berlin (1994).

12.9 PROBLEMS AND EXERCISES

1. What would be the effect of overlapping phosphor dots on a CRT screen?

2. An electron beam takes 0.1 s to return to the same phosphor dot on the screen of a CRT. If it is desired that the intensity of the excitation should not fall below one tenth of the initial intensity emitted, what must the decay time of the phosphor be?

3. A commercial colour TV phosphor has a decay time of 50 ns. If the fluorescence must fall to one hundredth of the initial value by the time the electron beam returns, how long will this take?

4. The band gap of gallium arsenide (GaAs) is 1.35 eV and the band gap of gallium phosphide (GaP) is 2.21 eV. What composition of a mixed crystal, $GaAs_xP_{1-x}$, would give orange light in an electroluminescent display?

5. Mercury vapour is not used in colour plasma displays. Why might this be?

CHAPTER 13

LASERS AND HOLOGRAMS

What are the most important characteristics of
laser light?
Why did lasers take so long to develop?
How was the population inversion problem solved in
the first ruby laser?
What are the benefits of neodymium (Nd^{3+}) solid
state lasers?
What lasers emitting coloured light are used to read
bar codes?
What lasers are used in CD players?

The word *laser* is an acronym for the expression *Light Amplification by Stimulated Emission of Radiation*. The first laser to be made was the ruby laser, and the first laser light emitted was on 15 May 1960. Since then a vast number of lasers have been produced, including solid-state lasers, gas lasers, semiconductor diode lasers and dye lasers. From an exotic beginning lasers have become ubiquitous in modern life, being used as pointers, at check-outs in supermarkets, in surveying and measurement, in micromachining, microsurgery and so on. Here the general principles of laser action will be outlined and some examples which illustrate particular facets of laser utility will be discussed.

13.1 EMISSION AND ABSORPTION OF RADIATION

When a photon of energy $h\nu$ is *absorbed* by an atom or molecule it passes from a lower energy state, called the *ground state*, to an upper or *excited* state, as described in earlier chapters and depicted in Figure 13.1. The transition will take place if the frequency of the photon, ν, is given exactly by:

$$\nu = (E_1 - E_0) / h$$

where E_0 is the energy of the ground state, E_1 is the energy of the excited state and h is Planck's constant.

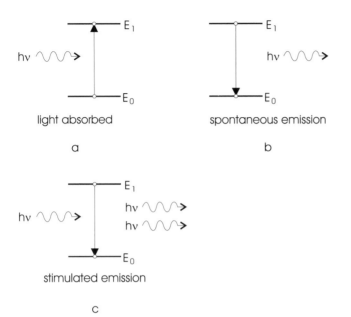

Figure 13.1 (a) Light absorption occurs when a photon excites an atom from the ground state, E_0 to an excited state E_1. (b) Light emission can take place by spontaneous emission if the atom loses energy at random. (c) Stimulated emission occurs when an atom in an excited state is triggered to lose energy by a passing photon of exactly the same energy (E_1–E_0)

If the atom is in the excited state, E_1, and makes a transition to the ground state, E_0, energy will be emitted with the same frequency, given by the same equation.

$$\nu = (E_1 - E_0) / h$$

Although the emission and absorption of radiation was briefly discussed earlier it is now necessary to look at these processes in a little more detail. In 1917 Einstein suggested that there should be *two* possible types of emission process rather than just one:

(i)   An atom in an excited state can randomly change to the ground state: a process called *spontaneous emission*.

(ii)   A photon having an energy equal to the energy difference between the two levels, that is, ($E_1 - E_0$), can interact with the atom in the excited state causing it to fall to the lower state and emit a photon at the same time: a process called *stimulated emission*.

These are illustrated in Figure 13.1. Lasers are concerned with stimulated emission.

In spontaneous emission, the light photons all have the same frequency but the waves possess random phases and the light is *incoherent*. In stimulated emission the photon produced has the same energy and frequency as the one which caused the emission and the light waves of both photons are *coherent*. It is these two important features of stimulated emission on which the special properties of laser light depend.

13.2 ENERGY LEVEL POPULATIONS

Under conditions of thermal equilibrium, the relative populations of a series of energy levels will be given by the Boltzmann law, which, for the case shown in Figure 13.1 can be written as:

$$(N_1/N_0) = \exp[-(E_1 - E_0)/kT]$$

where k is Boltzmann's constant, T is the absolute temperature, E_1 and E_0 are the energies of the excited state and the ground state, respectively, and N_1 and N_0 are the numbers of atoms (the *populations*) in each of these energy levels. Under normal circumstances the fraction N_1/N_0 will be extremely small for energy levels which are sufficiently separated to give rise to light.

To obtain laser action one needs to ensure that stimulated emission is the dominant process occurring. This means that a higher population of atoms in the upper level, E_1, than in the lower level, E_0, will be required. The reason for this is depicted in Figure 13.2. In Figure 13.2a the interaction of a photon with an atom in the ground state, E_0 is sketched. The photon will be absorbed and shortly afterwards re-released by spontaneous emission. This will be repeated at each atom in the ground state. There will be no amplification and we may well see a net absorption of energy. In Figure 13.2b a photon is shown interacting with an atom in the excited state. This time energy is released by stimulated emission and two photons emerge. If most atoms are in the excited state amplification will occur, as shown. The situation in which more atoms are in the excited state than the ground state is called a *population inversion*.

From the Boltzmann equation we can see that at normal temperatures we are very far from this situation. An increase in temperature cannot achieve this object either. Even at an infinite temperature one will only

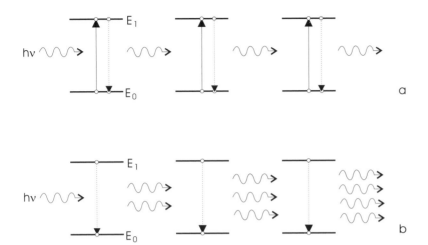

Figure 13.2 (a) When most atoms are in the ground state the absorption of a photon (full line) and its subsequent re-emission (dotted line) will not lead to amplification. (b) When most atoms are in the excited state stimulated emission rapidly leads to amplification

achieve equal numbers of atoms in E_0 and E_1. To obtain a population inversion, therefore, a *non-equilibrium state* must be achieved. The crux of laser action is how to create such a non-equilibrium situation in a material and then exploit it to produce the desired amplification.

13.3 RATES OF ABSORPTION AND EMISSION

In order to see how we can approach this problem it is necessary to consider the probabilities of spontaneous and stimulated emission and the probability of a photon being absorbed in more detail. First, we consider the equilibrium situation.

The rate of depopulation of an upper level ($-dN_1/dt$), by spontaneous emission will be given by:

$$-dN_1/dt = A_{10}N_1$$

where the negative sign denotes that the number N_1 of atoms in the upper state E_1 is decreasing with time. Not surprisingly, this tells us that the rate is proportional to the number of atoms in the state, N_1. The rate constant, denoted here as A_{10}, is called the *Einstein coefficient for spontaneous emission*, and the suffix 10 means that we are considering a transition from the excited state, E_1 to the ground state E_0. The number of downward transitions due to spontaneous emission, per second will be given by:

$$A_{10}N_1$$

In similar fashion, two other rates can be defined for the cases of stimulated emission and for absorption. These can be expressed in terms of two rate constants (or Einstein coefficients), one for stimulated emission and one for absorption. As expected, the rates are proportional to the numbers of atoms in the relevant state and the number of photons present. Thus the rate at which atoms in state E_0 are excited to state E_1 is then given by:

$$-dN_0/dt = B_{01}\rho(\nu_{01})N_0$$

where N_0 is the number of atoms in state E_0 per cubic metre, $\rho(\nu_{01})$ is the radiation density responsible for absorption, which is the number of quanta per cubic metre incident per second at the correct excitation frequency ν_{01}, and B_{01} is the *Einstein coefficient for absorption of radiation*. Similarly, the rate of depopulation of state E_1 by stimulated emission is given by:

$$-dN_1/dt = B_{10}\rho(\nu_{10})N_1$$

where N_1 is the number of atoms in state E_1 per cubic metre, $\rho(\nu_{10})$ is the radiation density responsible for depopulation, which is the number of quanta per cubic metre incident per second at the correct frequency ν_{10}, and B_{10} is the *Einstein coefficient for stimulated emission of radiation*. Now the correct frequency for excitation will be the same as that for depopulation, so that $\nu_{10} = \nu_{01}$, which we can simply write as ν, and the radiation density will be the same in each case, so that we can write:

$$\rho(\nu_{10}) = \rho(\nu_{01}) = \rho(\nu)$$

The number of stimulated downward transitions per second will be given by:

$$N_1 B_{10}\rho(\nu)$$

while the total number of upward transitions in the same time will be given by:

$$N_0 B_{01} \rho(\nu)$$

At equilibrium, the total number of transitions in each direction must be equal, hence:

$$N_0 B_{01} \rho(\nu) = N_1 A_{10} + N_1 B_{10} \rho(\nu)$$

so

$$\rho(\nu) = N_1 A_{10} / (N_0 B_{01} - N_1 B_{10})$$

At equilibrium the Boltzmann distribution applies, thus:

$$N_1/N_0 = \exp(-h\nu/kT)$$

and by making this substitution:

$$\rho(\nu) = A_{10} / [\exp(h\nu/kT)B_{01} - B_{10}]$$

This expression represents the net radiation emitted or absorbed by the material. Comparing this with Planck's equation for black-body emission in Chapter 1:

$$\rho(\nu) = 8\pi h\nu^3/[c^3 \exp(h\nu/kT) - 1]$$

leads to the conclusion that:

$$B_{01} = B_{10}$$

which will be replaced by the single symbol B, and

$$A_{10}/B = 8\pi h\nu^3/c^3$$

The *ratio* of the rate of spontaneous emission to stimulated emission under conditions of thermal equilibrium, given by:

$$R = A_{10} / \rho(\nu)B = \exp(h\nu/kT) - 1$$

This is an extremely interesting result. At 300 K, at visible wavelengths, R is much greater than 1. This shows that for light, stimulated emission will be negligible compared to spontaneous emission. Incidentally it ought to mean that it will be impossible to make a laser! On the other hand, if the wavelength increases beyond the infrared into the microwave and radio wave regions of the electromagnetic spectrum, R becomes *much less* than 1 and *all* emission will be stimulated. Hence, radiowaves and microwaves arise almost entirely from stimulated emission and are always coherent. This is one of the main reasons that communications in the early part of the 20th century used radio waves.

The following sections show how the equilibrium problem has been bypassed and how the difficulty of achieving stimulated emission at optical wavelengths has been overcome.

13.4 THE RUBY LASER: A THREE-LEVEL LASER

As we have seen, for laser action two objectives have to be fulfilled. It is necessary to obtain a population inversion and to ensure that the higher energy level is depopulated by stimulated emission and not by spontaneous emission. This was first achieved in the ruby laser. In order to see how this was possible we need to consider the energy level diagram relevant to the transitions taking place in this material.

The electron transitions which lead to colour in rubies have been discussed in Chapter 7. It is necessary to expand the previous discussion a little in order to explain the production of a population inversion. Optical transitions are constrained by quantum theory, which allows one to calculate the probability that a transition will take place. Those which are highly probable are said to be *allowed* transitions while those which are unlikely are said to be *forbidden*. Remember, though, that because we only assess probabilities, forbidden transitions still have a low probability of occurring. One of the rules that the calculations lead to is that optical transitions are only allowed between levels in which the total amount of electron spin does not change. (It is only the way in which these spins have combined in the atom or ion that must not change, not the spin on any one electron, which is constant at 1/2.) This quantity is called the *spin multiplicity* of the energy state. Such transitions are called *spin allowed* transitions. In the case of ruby there are two important spin allowed transitions. With reference to Figure 13.3, these are:

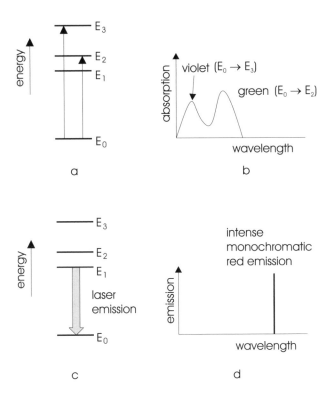

Figure 13.3 The transitions of the Cr^{3+} ion important in the laser action of ruby. (a) The spin-allowed transitions which give colour to the ruby in normal light. (b) The absorption spectrum of a ruby crystal. (c) In laser action, optical pumping to the E_2 and E_3 levels is followed by radiationless transitions which allows energy to be lost to the crystal structure and the Cr^{3+} ions to enter the state E_1. Transitions from E_1 to the ground state, E_0, are not spin allowed and laser emission takes place by the stimulated return to the ground state. (d) The laser emission is monochromatic red light at a wavelength of 694 nm in the red

$E_0 \rightarrow E_2$ (at 556 nm, absorbs yellow/green)
$E_0 \rightarrow E_3$ (at 407 nm, absorbs violet).

Now the third level shown in Figure 13.3, level E_1, has a different total spin to the levels E_2 and E_3, and normally would not be involved in the optical transitions to any great extent. However, in ruby, excited Cr^{3+} ions in states E_2 or E_3 can lose energy to the crystal lattice. This process operates under different conditions than the optical transitions and is independent of spin. Thus it is possible for Cr^{3+} ions in states E_2 or E_3 to lose energy to the lattice and drop down to level E_1. In this transition the energy is taken up in lattice vibrations and the ruby crystal warms up. This is called a *radiationless* or *phonon assisted transition*. Typical rates of the transitions are:

$$E_2 \rightarrow E_0 \quad 3 \times 10^5 \text{ s}^{-1}$$
$$E_2 \rightarrow E_1 \quad 2 \times 10^7 \text{ s}^{-1}$$

The second of these two transitions is about 100 times faster than the first. The rates of the transitions from the E_3 energy level to E_1 and E_0 are of a similar magnitude. This means that on irradiating the ruby with white light, a significant number of atoms end up in the E_1 state. Now the transition from E_1 to the ground state is not allowed because of the spin rule and so atoms in the E_1 state have a long lifetime. (The spontaneous emission rate is $2 \times 10^2 \text{ s}^{-1}$.) Thus it is possible to build a population inversion between the E_1 and E_0 levels.

Laser operation takes place in the following way. An intense flash of white light is directed onto the crystal. This process is called *optical pumping*. This excites the Cr^{3+} ions into the E_2 and E_3 states. These then lose energy by radiationless transitions and "flow over" into state E_1. If the initial flash is intense enough this will cause a population inversion between E_1 and E_0. About 0.5 ms after the start of the pumping flash, some spontaneous emission will occur from E_1. In order to prevent these first photons from escaping from the crystal without causing stimulated emission from the other excited ions one end is coated with a mirror and the other with a partly reflecting mirror. In this case the photons are reflected to and fro, causing stimulated emission from the other populated E_1 levels. Once started, the stimulated emission rapidly depopulates these levels in an avalanche. There will be a burst of red laser light of wavelength 694.3 nm which emerges from the partly reflecting surface. In the original laser silver mirrors were used. However, silver absorbs as much light as it transmits. This resulted in overheating which could cause the crystals to shatter. Now the mirrors are thin film dielectric mirrors of the sort described in Chapter 4, one of which transmits about 1 per cent of the incident photons.

Following the light burst, the upper levels will be empty and the process can be repeated. The ruby laser generally operates by emitting energy in short bursts, each of which lasts about 1 ms. This is referred to as *pulsed* operation.

The ruby laser is called a *three-level laser*, because basically three energy levels are involved in the operation. These are the ground state (E_0), an excited state reached by optical absorption or pumping (E_2 or E_3), and an intermediate state of long life-time (E_1) reached by radiationless transfer from the previous state and from which stimulated emission (laser emission) occurs to the ground state. It is energetically costly to obtain a population inversion in a three-level laser because one must pump more than half the population of the ground state to the middle level. Moreover, very little of the electrical energy supplied to the flash lamp ends up pumping photons, and carefully designed reflectors are essential. Finally, the energy lost in the transitions from E_3 and E_2 to E_1 ends up as lattice vibrations, which cause the crystal to heat up considerably. To make sure that the ruby does not overheat and shatter it is necessary to cool the crystal and to space the pulses to allow the heat to dissipate.

13.5 THE NEODYMIUM (Nd^{3+}) SOLID-STATE LASER: A FOUR-LEVEL LASER

Although the ruby laser was the first laser made, the three-level operation makes it inefficient. A more energy-efficient device can be made employing an energy level scheme similar to that shown in Figure 13.4. Lasers using this type of energy level arrangement are referred to as *four-level lasers*.

Laser operation takes place in the following sequence of steps.

(i) Atoms in the ground state E_0 are excited to a rather high energy level, E_2 by optical pumping. This process needs to be fast and efficient.

(ii) The atoms in E_2 lose energy without radiating light, to an intermediate state I_1. This step should also be fast and efficient. However, once in I_1, atoms should have a long lifetime and not lose energy quickly.

(iii) It is essential that another intermediate state, I_0, is present and also sufficiently high above the ground state to be effectively empty. In this case a small population in I_1 gives a population inversion between I_1 and I_0.

(iv) Ultimately a few photons will be released as some atoms drop from I_1 to I_0. These can promote stimulated emission between I_1 and I_0, allowing laser action to take place.

(v) Atoms return from I_0 to E_0 by a step which needs to be rapid.

(vi) If the energy corresponding to the transitions from E_2 to I_1 and I_0 to E_0 can be easily dissipated, continuous operation rather than pulsed operation is possible.

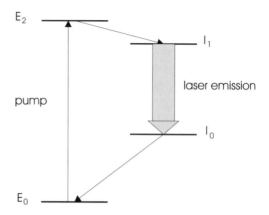

Figure 13.4 Principle of action of a four level laser. The laser is initially pumped to promote the active atoms or ions from the normally occupied ground state, E_0, to an excited state E_2. The excited state E_2 loses energy so as to populate an intermediate state I_1. Laser action can take place between I_1 and another intermediate state I_0, provided that the population of I_1 is greater than I_0. Finally the population in I_0 decreases as the atoms or ions return to the ground state E_0 by energy loss, allowing the cycle to be repeated

The most important four-level solid-state laser uses neodymium (Nd^{3+} ions) as the active centres. As we saw in Chapter 7, lanthanide ions can be introduced into a wide variety of host lattices with little effect on optical properties because the important 4f-orbitals are shielded from the crystal surroundings by outer 5d- and 6s-electron orbitals. This allows for the use of a wide variety of host lattices for the Nd^{3+} ions while still retaining the energy level scheme necessary for laser action. The most common host materials are glass, yttrium aluminium garnet (YAG) and calcium tungstate ($CaWO_4$).

The important transitions taking place in Nd^{3+} ion lasers can be understood in terms of the simplified energy level diagram drawn in Figure 13.5. The energy levels arise from the f-electron orbitals and outer orbitals. The f-electron levels are rather sharp. Above these lie bands of considerable width derived from the interaction of the 5d- and 6s-orbitals. Optical pumping excites the ions from the ground state to these wide bands. This process is very efficient because broad bands allow a wide range of wavelengths to pump the laser and because the transitions are allowed in terms of quantum theory. In addition, loss of energy from the excited state down to the f-electron energy levels is fast. The energy loss halts at the pair of levels labelled ^4F in Figure 13.5.

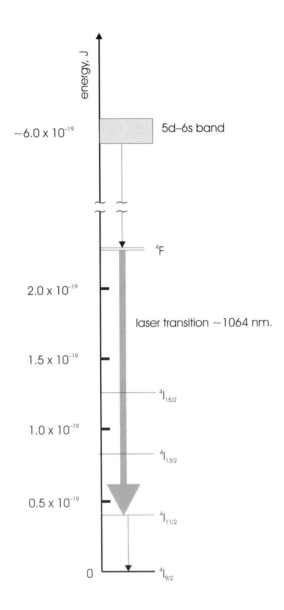

Figure 13.5 Energy levels for an Nd³⁺ ion in a laser host material. At the highest energies there is a broad 5d–6s band. At lower energies a series of sharp energy levels exist. On the figure these are drawn as single lines, but in reality each of these levels is comprised of a number of closely spaced levels. The principal laser transition is shown as an arrow between the ⁴F and ⁴I₁₁/₂ levels

The principal laser transition is from these ^{4}F levels to a set of levels labelled ^{4}I$_{11/2}$. The emission is at approximately 1064 nm in the infrared. This is a useful wavelength as it coincides with a reasonably low-loss region of silica based optical fibres shown in Figures 11.3 and 11.5. The lasers contain about 1% Nd^{3+} and can have quite high power outputs. They can be operated continuously or pulsed.

At higher Nd^{3+} concentrations the lifetime of the ^{4}F upper state drops from about 200 μs in a typically 1% doped material to about 5 μs at higher dopant concentrations. This is due to Nd–Nd interactions and associated changes in lattice vibration characteristics. Under these conditions laser operation is no longer possible.

13.6 THE HELIUM–NEON LASER

The first four-level laser actually developed was the helium–neon (He–Ne) laser, which was originally demonstrated in 1960. This laser is the ubiquitous red laser common in supermarket check-out counters and laser pointers, although colours other than red can also be produced by the helium–neon combination and the first laser wavelength produced was, in fact, at 1.15 μm. The laser consists of a low pressure (10^{-2} to 10^{-3} atm.) of helium mixed with about 10% neon, enclosed in a narrow glass tube. High energy electrons are produced by subjecting the cathode (the negative terminal) to a high voltage, as in a normal fluorescent tube light. These energetic electrons, e^{-*}, collide with the helium atoms to produce an excited state, He*:

$$He\ (1s^2) + e^{-*} \rightarrow He^*\ (1s^1\ 2s^1) + e^-$$

The $1s^1 2s^1$ configuration gives rise to two energy levels with term symbols 2^1S and 2^3S, as indicated on Figure 13.6. The excited He* can pass its energy over to a neon atom during a collision to produce an excited neon atom, Ne*. This can happen because, quite by chance, the energy transferred almost exactly matches an excitation energy of Ne, as shown in Figure 13.6.

$$He^*\ (2^1S) + Ne\ (2p^6) \rightarrow He + Ne^*\ (2p^5\ 5s^1)$$
$$He^*\ (2^3S) + Ne\ (2p^6) \rightarrow He + Ne^*\ (2p^5\ 4s^1)$$

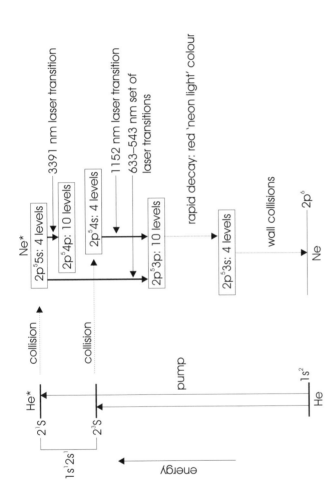

Figure 13.6 Schematic processes operating in a helium–neon laser. Helium (He) atoms are excited from the $1s^2$ ground state into the two $1s^1 2s^1$ (He*) states and transfer energy to neon (Ne) atoms to excite them into the $2p^5 4s$ and $2p^5 5s$ states. The main laser transition is from the $2p^5 5s$ state to the $2p^5 3p$ state, from which the atoms return in two steps to the $2p^6$ ground state

Table 13.1 Energy levels in neon relevant to laser action

Electron configuration	Laser terminology
$2p^5 3s^1$	$1s_2$–$1s_5$ (4 energy levels)
$2p^5 3p^1$	$2p_1$–$2p_{10}$ (10 energy levels)
$2p^5 4s^1$	$2s_2$–$2s_5$ (4 energy levels)
$2p^5 4p^1$	$3p_1$–$3p_{10}$ (10 energy levels)
$2p^5 5s^1$	$3s_2$–$3s_5$ (4 energy levels)

The neon energy levels derived from these two configurations each consist of four levels. In addition there are two other important sets of 10 levels present on the neon atoms, derived from the configurations $2p^5 4p^1$ and $2p^5 3p^1$. All of the energy levels on the neon atoms are complex and the term symbols found in texts on atomic spectra are obscure. Laser workers therefore use a simpler labelling system of s and p designations, given in Table 13.1. Unfortunately this mimics the chemical configuration symbols explained in Appendix 7.1, but without the same implications, which leads to unnecessary confusion. In order to relate Figure 13.6 to the labels found in laser texts the correspondence in nomenclature is given in Table 13.1.

As indicated above, the collisions between neon and excited helium atoms (He*) produces a population of excited neon atoms (Ne*) in which several series of occupied and empty energy levels exist in close conjunction. These excited Ne* atoms can release energy by stimulated emission thereby dropping to many of these empty levels, and about 100 or more output wavelengths can appear. The main transition, however, is from the highest $2p^5 5s^1$ set of levels to the $2p^5 3p^1$ set of levels,

$$\text{Ne}^* (2p^5 \, 5s^1) \rightarrow \text{Ne}^* (2p^5 \, 3p^1) + h\nu$$

The transition produces the well known red laser output with a wavelength of 632.8 nm. Transitions to some of the other levels in the same manifold give rise to the other coloured output frequencies, which include 543.5 nm (green), 594.1 nm (yellow) and 612.0 nm (orange).

The still energised neon atom thereafter rapidly decays to the ground state, $2p^6$ in two steps.

$$\text{Ne}^* (2p^5 \, 3p^1) \rightarrow \text{Ne}^* (2p^5 \, 3s^1) \rightarrow \text{Ne} (2p^6)$$

The $2p^53p^1$ to $2p^53s^1$ transition is fast and helps to maintain a population inversion between the $2p^53p^1$ level and those above it. This transition is responsible for the red neon light colour discussed in Chapter 7. The final transition is radiationless, and energy is often lost to the walls of the laser tube in transitions which do not give out light. At this stage the process can begin all over again.

13.7 Semiconductor Diode Lasers

If a forward bias voltage is applied to a p-n junction it will function as a light emitting diode or LED (see Chapter 12). In principle it is very easy to turn an LED into a semiconductor laser, a *diode laser*. As with other lasers, it is necessary to obtain a population inversion and then to activate stimulated emission.

A population inversion is achieved by using a heavily doped n-type region with respect to the p-type region. It comes about in the following way. When a low forward bias is applied to the diode, a small and roughly equal number of holes and electrons enter the junction region, recombine and emit light. This is normal LED behaviour, and produces light more or less in proportion to the magnitude of the current. However, at some critical threshold voltage far more electrons start to enter the junction region than holes. This is because the hole transport into the junction region is at a plateau, the height of which is controlled by the relatively low doping level. However, the electron transport into the junction region can continue to rise because the n-type segment is heavily doped and has a far higher plateau. At the point when the electron numbers outweigh the number of holes a population inversion occurs and the LED becomes a laser. Stimulated emission is achieved by using carefully polished crystals so that any photon emitted will be reflected to and fro in the junction to promote the laser avalanche.

The change from LED to laser operation is marked by a large increase in both output and efficiency. The change is shown schematically in Figure 13.7a. The light is emitted from the small junction region as shown in Figure 13.7b. Because of this semiconductor lasers can be small and are often comparable in size to a grain of salt.

Diode lasers are most often made from materials related to gallium arsenide (GaAs) or indium phosphide (InP). The output wavelengths fall into the ranges of approximately 630–980 nm for GaAs derived systems and 1300–1550 nm for InP derived systems. In practice simple diodes

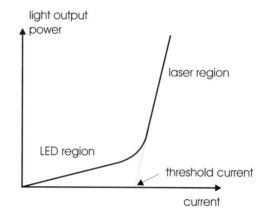

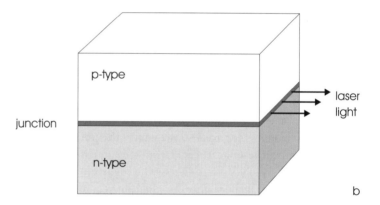

Figure 13.7 A p-n junction laser. (a) A positive voltage above the threshold voltage applied to the p-type region of an LED transforms it into a semiconductor diode laser. (b) Light is emitted from the junction region in a narrow beam

consisting of an n-type and p-type region in juxtaposition are not used. Instead an arrangement called a *double heterojunction* laser is employed. This device is a laser consisting of a thin layer of undoped gallium arsenide (called a well) which separates p-type and n-type gallium aluminium arsenide regions. Photons are emitted in the junction region but are confined to the well due to refractive index differences between the gallium arsenide and the gallium aluminium arsenide. This confinement enhances stimulated emission and gives a more collimated beam. Such heterojunction lasers are widely used as the light sources for optical fibre communications and as the reading source for compact discs.

Compact discs (CDs) and the more recent digital video discs (DVDs) store the information digitally. The principle of encoding a sound wave digitally was outlined in Chapter 11. In the case of discs, the 0s and 1 s are represented by a series of microscopic pits etched into the surface of the disc. This is made of a polycarbonate plastic and coated with a thin reflective film of aluminium. To play the disc the surface is spun under the beam from a diode laser. The reflectivity of the surface depends upon whether a pit lies under the beam at any moment. Differences in reflectivity are converted firstly into 0s and 1s and then into the sound produced by the speakers.

Information storage capacity will depend upon the size of the pits and the size of the scanning light beam. Present day technology uses mainly infrared diode lasers as these have greatest reliability of operation. However, there is much research into diode lasers which can emit light in the blue to violet region of the spectrum or even into the ultraviolet. The reason for this is that the amount of information written on a conventional compact disc could be increased fourfold by virtue of the shorter wavelength of blue light. A subsidiary benefit of shorter wavelengths is the higher energy of the photons. This higher energy would usefully shorten writing time. At present most hope is pinned on gallium nitride (GaN) as the base material for a blue laser or diode, as this has been demonstrated to give out short wavelength light. However, at the moment, commercial devices using GaN lasers are only just becoming available.

13.8 DYE LASERS

Dye lasers differ from the solid state and gas lasers described in a significant way. The output can be tuned over a wide range of wavelengths. In the other lasers mentioned the energy levels between which the laser transitions occurred were discrete and the output is a series of sharp lines. In order to alter the output one has to use frequency doubling or tripling, optical parametric oscillators or up-conversion. Molecules have rather broad energy levels which are better described as bands, due to the addition of vibrational and rotational energy levels to each electronic level, as was noted in Chapter 8. Dye molecules are large and although rotational levels are not important vibrational levels remain so, as illustrated in Figure 13.8.

When a dye molecule is excited an electron moves from the lower HOMO (symbol S_0) to the upper LUMO (symbol S_1), as indicated in Figure 13.8a. Both of these states have associated vibrational levels, indicated in Figure

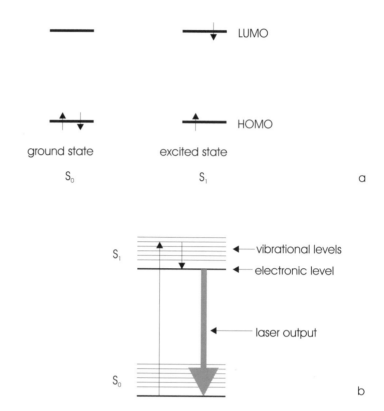

Figure 13.8 (a) The ground state, S_0, and the excited state, S_1, of a typical dye molecule. (b) Excitation of the molecule promotes an electron from the ground state to the excited state. The molecule loses energy via collisions to reach the bottom level in the S_1 set before a laser transition returns it to one of the vibrational levels in the ground state

13.8b. The absorption spectrum is broad because the excitation can take the molecule from the ground state into many of the vibrational levels associated with the upper energy level. Energy is rapidly lost, by collisions with other molecules and the molecule, rapidly ends in the lowest level of the excited state. Laser action can now occur when the molecule drops to any of the empty vibrational energy levels of the ground state, as drawn in Figure 13.8b. Like the absorption spectrum, the emission spectrum is broad because of the number of vibrational levels, and it is also displaced slightly with respect to the absorption spectrum due to the loss of energy as the excited state decays to its lowest level. The absorption and emission spectra of the dye rhodamine 6G, used in dye lasers, is drawn in Figure 13.9.

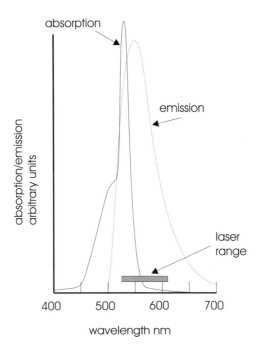

Figure 13.9 Schematic diagram of the absorption and emission spectra of the dye rhodamine 6G. The useful range of laser output for a dye laser using this molecule is indicated

In practice many dye molecules can be used but those with efficient fluorescence are naturally preferred. In use the dye is dissolved in a suitable solvent, often methanol or ethanol. The energy loss as the excited molecules decay through the vibrational levels is transmitted to the solution as heat, which can seriously impair the performance of the laser. To avoid this, the dye solution flows continuously through the glass cell in which the laser action takes place from a temperature-controlled bath, so as to keep the solution at the optimum temperature. Optical pumping of the dye molecules usually makes use of another laser and mirrors enclosing the cell ensure that efficient stimulated emission occurs. If power levels and dye flow rates are adjusted, dye lasers can operate in a continuous mode as well as a pulsed mode. A practical arrangement is schematically depicted in Figure 13.10. The strong absorption of dye molecules allows the laser cell to be small and a path length of several centimetres will suffice in the majority of cases. As the emission spectrum is broad, the output wavelength can be selected using a diffraction grating as a tuner. The multiplicity of dyes

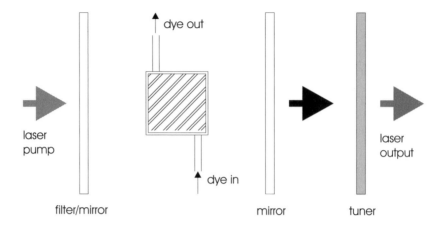

Figure 13.10 Schematic diagram of a dye laser. The filter/mirror passes pump radiation but reflects dye emission. The dye solution flows into and out of the cell in which the laser emission takes place from a temperature-controlled reservoir. The mirror on the output side partially reflects the dye emission. The tuner is frequently a diffraction grating which allows a narrow range of output wavelengths to be selected from the broad emission spectrum

Table 13.2 Some dye molecules used in lasers

Dye	Output range (nm)
Coumarin 9	430–530
Rhodamine 6G	540–605
Rhodamine B	580–655
Oxazine 9	644–709

available means that the whole of the visible spectrum is easily accessible. Some of the commoner dyes used are listed in Table 13.2.

13.9 HOLOGRAMS

Objects are perceived when a light wave enters the eye and the resulting nerve impulses are processed by the brain. Strictly speaking the source of the wavefront entering the eye is unimportant for the perception to occur. Holograms are means of creating wavefronts in such detail that the observer is given the impression that the real object is being observed even though it is, in fact, an optical illusion. Holograms are, in fact, two- or three-dimensional interference patterns, which, when illuminated with the correct light, recreate the wavefront originally derived from the object. This generates a virtual image which is produced by an exact copy of the wavefront originally by the reflected object.

A schematic idea of the formation of a hologram is given in Figure 13.11a. A monochromatic laser beam with a wavelength somewhere in the visible is divided into two as shown. One part illuminates an object and some of the light reflected from the object falls onto a photographic film. The other part of the beam traverses an identical distance and also arrives at the photographic film, but without having encountered the object. At this point the two reunited beams will interfere with each other and the interference pattern will be recorded in the photographic emulsion. This interference pattern is thus a pattern of coded information which records the changes made to that part of the beam which was reflected from the object. The interference pattern is made permanent by the usual photographic processing steps. If the hologram is now illuminated by an identical laser and the hologram is viewed in *transmission*, as shown in Figure 13.11b, the wavefront received by the eye will be identical to the wavefront that the eye would have received after reflection from the object itself. Should the observer move, a different portion of the wavefront will be intercepted by the eye and this will generate a fully three-dimensional effect identical to that which would be experienced if the original object were present.

There are two drawbacks to this procedure. First, in order to make such a hologram all vibrations must be eliminated. Any disturbance at all will introduce additional "information" into the hologram, which will be interpreted by the eye as a degradation of the image. Second, it is necessary to view the hologram with laser light of the same wavelength as was used in its creation in order to perceive the effect.

These difficulties can be removed and holograms can be made and viewed in white light by a simple modification of the procedure outlined. In this case the object is actually clamped to the back of a photographic film, as depicted in Figure 13.12a. Because the photographic emulsion and the object are fixed together vibrations will affect both together and not introduce any new information. The photographic emulsion does not absorb all of the incoming photons and some pass through to be reflected by the object behind the film. A three-dimensional interference pattern is now created within the emulsion by interaction between the incoming wavefront and the reflected wave from the object. It is usual to bleach the emulsion after it has been processed so as to remove the silver particles present. In this case the interference pattern is recorded as an arrangement of varying refractive index rather than of differing densities of silver

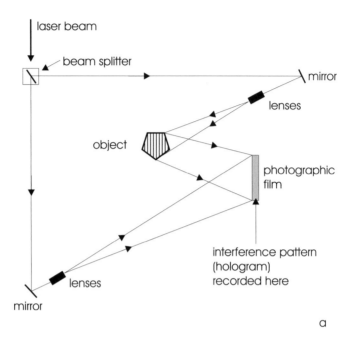

a

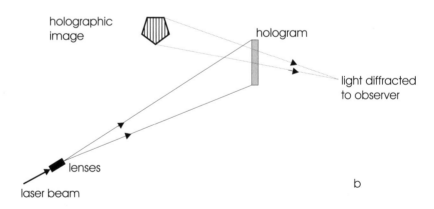

b

Figure 13.11 The principle of making a hologram. (a) An object is irradiated with part of the beam from a laser. The light reflected from the object interferes with the second part of the beam within the photographic film to create an interference pattern (hologram) in the emulsion. (b) When the processed film is viewed in transmission in the same laser light the eye sees an exact reconstruction of the object

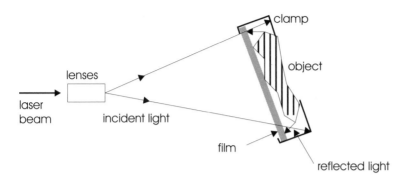

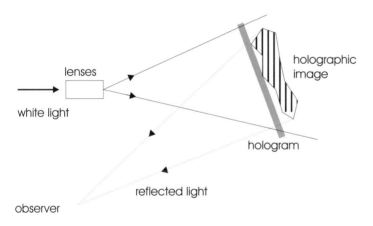

Figure 13.12 The principle of making a reflection hologram. (a) Laser light is reflected from an object in contact with a photographic film. This reflected light creates an interference pattern within the film by interaction with the incoming beam. (b) Viewing the film in reflected white light will recreate an apparent copy of the object. Because different wavelengths of white light interact differently with the hologram, the image will often appear to be coloured

particles. The result is a *reflection* hologram. When such a hologram is viewed in white light, as indicated in Figure 13.12b, at least one wavelength present in the light will match that of the source laser and a three dimensional reconstruction will be seen in this colour. A slight tilting of the hologram will allow a different wavelength to interact and the image appearance and colour will change accordingly. Because of the rather complex three dimensional pattern of refractive index variation recorded other interference effects can also occur and variously coloured holographic images will be seen as the viewing angle is changed. This effect gives rise to the commonplace coloured holographic images now used as logos in advertising and on credit cards. In these the hologram is contained in a thin plastic film which is mounted on the reflecting surface of a piece of aluminium foil. The high reflectivity of the foil ensures that the colours are readily visible even in bright daylight while the holographic nature of the information recording means that the image changes and gives different three-dimensional impressions as the object is tilted.

It is clear that a hologram contains as much information as a collection of photographs taken from different angles. In this way a hologram is a very efficient way of storing information about objects. Holograms are permanent and can be retrieved easily. Indeed, in many instances, such as when viewing a transient phenomenon, they may be preferable to viewing the original object itself. Holograms of a surgical procedure, for example, can give full information without the necessity of students or others having to be present at the operation.

13.10 ANSWERS TO INTRODUCTORY QUESTIONS

What are the most important characteristics of laser light?
The most important characteristic of laser light is that it is coherent. Other very useful attributes are that it is very nearly monochromatic and it is possible to obtain light pulses of very short duration (10^{-12} s or less).

Why did lasers take so long to develop?
One important reason for this was that the Einstein theory of absorption and emission of radiation suggested that it was not possible to obtain a population inversion for energy levels of importance to light emission.

How was the population inversion problem solved in the first ruby laser?
The population inversion was achieved by making use of a very intense pumping flash of light and the fact that the excited state of the Cr^{3+} ions which was produced by the flash decayed to a state with a very long lifetime by losing energy to the crystal lattice.

What are the benefits of neodymium (Nd^{3+}) solid state lasers?
Nd^{3+} lasers offer several advantages over ruby lasers. The laser operation is of the four-level type and so much more efficient. The laser can be operated

to give a continuous wave output and because neodymium is a lanthanide the emission is almost the same irrespective of the host lattice used.

What lasers emitting coloured light are used to read bar codes?
These are helium–neon gas lasers.

What lasers are used in CD players?
These are very small diode lasers which may operate in the visible or infrared. They are often made using doped gallium arsenide.

13.11 FURTHER READING

The invention of the first laser is reported by:

T. H. Maiman Nature, **187**, 1960, p. 493.

Lasers are dealt with from an electronic engineering point of view in:

L. Solymar and D. Walsh, Lectures on the Electrical Properties of Materials, 5th edition, Oxford University Press, Oxford (1993), Ch. 12.

Detail is given in:

A. E. Siegman, Lasers, University Science Books, Sausalito, California (1986).
O. S. Heavens and R. W. Ditchburn, Insight into Optics, Wiley, New York (1991), Ch. 13.

Lasers are discussed from an experimental point of view in:

R. N. Zare, B. H. Spencer, D. S. Springer and M. P. Jacobson, Laser Experiments for Beginners, University Science Books, Sausalito, California (1995).

Holography from a practical standpoint is covered by:

J. Walker, Scientific American, **260**, May 1989, p. 100 and references therein.

It is covered from both a practical and theoretical point of view in:

F. Unterseher, J. Hansen and B. Schlesinger, Holography Handbook, Ross Books, Berkeley (1996).
J. E. Kasper and S. A. Feller, The Complete Book of Holograms and how to make them, Wiley, New York (1987).

13.12 PROBLEMS AND EXERCISES

1. Estimate the fraction of Cr^{3+} ions atoms in the upper (E_1) state for a ruby laser when populated thermally at 300 K. The laser transition in ruby is in the red part of the visible spectrum and the emission occurs at a wavelength of 694.3 nm.

2. Why can population inversion between the levels E_1 and E_0 of Q1 not be obtained thermally?

3. At what wavelength and frequency does stimulated emission become more important than spontaneous emission?

4. Estimate the equilibrium ratio of spontaneous to stimulated emission for a hypothetical laser operating with an output wavelength of 556 nm.

5. What type of radiation process, spontaneous or stimulated, dominates microwave emission of radiation?

6. What is the power of a laser emitting 100 J in 1 ms?

7. What is the separation of the energy levels in neodymium ions (Nd^{3+}) which give rise to the emissions at 0.914 μm and 1.06 μm?

8. Explain why solid-state lasers using lanthanide ions employ those with few f electrons, typically Nd^{3+}, while for up-conversion and amplification (see Chapter 11) those with many f electrons, such as Er^{3+}, are preferred.

ANSWERS TO NUMERICAL PROBLEMS

Full annotated answers to all of the problems and exercises will be found at http//www.cardiff.ac.uk/uwcc/engin/staff/rjdt/colour.

CHAPTER 1

4. 425 nm = 7.6×10^{-14} Hz = 4.68×10^{-19} J (violet)
    575 nm = 5.22×10^{-14} Hz = 3.46×10^{-19} J (yellow)
    630 nm = 4.76×10^{-14} Hz = 3.15×10^{-19} J (red)

9. $\alpha = 0.95$ m^{-1}

CHAPTER 2

1. If the glass is in air, $\theta_c = 27.6°$
    If the glass is coated, $\theta_c = 37.7°$

2. n = 1.71 (observed, 1.72).

3. n = 1.560 (observed, 1.586).

4. $\rho = 2.93$ g cm^{-3}.

5. k = 0.20.

6. n = 2.646.

7. $v_d = 64.2$ (the measured value is 64.1).

8. 0.3 mm (the red band is at the top).

CHAPTER 3

1. 53°.

2. $R_s = 0.05 = 5\%$.
    $R_p = 0.04 = 4\%$.

3. Fifty per cent of the incident light will be transmitted.

5. Scheelite, 0.016; corundum, 0.009; forsterite, 0.035; malachite; 0.252; sterconite, 0.030.

6. For an angle of 20°, $n_e' = 1.545$.
    For an angle of 45°, $n_e' = 1.5485$, which is the mean of the two principal refractive indices.

7. The retardation of the foil is 505 or 998 nm.
    The thickness of the foil is 13 μm or 25.6 μm.

CHAPTER 4

1. 5.3×10^{-2} or 5.3%.

4. 47.4 nm; yellow-white to straw yellow for reflection and carmine-red to deep violet for transmission.

5. $R = 0.148 = 14.8\%$.

8. The new retardation is 1191 nm. The colour reflected will correspond to third order blue with a green tinge.

9. Second order orange-red.

12. $R = 0.013 = 1.3\%$.
    (This will only apply to one particular wavelength, the *design wavelength*, because the refractive indices vary from one wavelength to another.)

13. $R = 0.477 = 47.7\%$.

14. If $N = 1$, $R = 0.465 = 46.5\%$.
    If $N = 2$, $R = 0.8071 = 80.7\%$.
    A continuation of the calculation shows that R rapidly nears 100%.

CHAPTER 5

1. $\lambda = 400$, $R = 1.000$ (100%).
    $\lambda = 450$, $R = 0.625$ (62.5%).
    $\lambda = 500$, $R = 0.410$ (41.0%).

λ = 550, R = 0.280 (28.0%).
λ = 600, R = 0.198 (19.8%).
λ = 650, R = 0.143 (14.3%).
λ = 700, R = 0.107 (10.7%).

4. Scattering becomes as important as absorption when 1.6 times the particle radius is equal to the wavelength of the light.

5. An increase in length/width would lead to yellow, orange and then red.

CHAPTER 6

1. w = 9.5×10^{-5} m.

2. 0.004 mm.

4. 0.1 mm.

5. 0.4177 nm.

6. 3.2 nm.

10. 175 nm.

11. 201.6 nm.

CHAPTER 8

2. $\lambda \sim$ 385 nm for $CH_3(CH=CH)_7CH_3$ and 420 nm for $CH_3(CH=CH)_9CH_3$. The first of these would be colourless unless the band encroached into the visible, in which case it would absorb some violet and look weak yellow-red. The second would absorb strongly in the indigo region and appear strongly orange-yellow.

5. 3.75.

CHAPTER 9

6. 4.580×10^{-19} J (275.7 kJ mol^{-1}).

CHAPTER 10

1. Mercury Hg, R = 77.9%.
   Chromium Cr, R = 65.5%.
   Titanium Ti, R = 50.9%.

6. $Na_{0.25}K_{0.75}Br$.

CHAPTER 11

1. 3.0 dB km^{-1}.

2. 6 cm.

3. 49 910.638 m; 2.2 ns km^{-1}.

4. 200 km.

5. 6.7 ns km^{-1}.

CHAPTER 12

2. 0.43 s.

3. 2.3×10^{-7} s.

4. $GaAs_{0.29}P_{0.71}$.

CHAPTER 13

1. 9.95×10^{-31}.

3. ν, 4.33×10^{12} s^{-1} (Hz); λ, 6.92×10^4 nm.
   This is in the infrared.

4. 3.2×10^{37}.
   At 300 K stimulated emission will be negligible compared to spontaneous emission.

6. 10^5 W.

7. 914 nm, 2.17×10^{-19} J; 1060 nm, 1.87×10^{-19} J.

Formula Index

SUBJECT INDEX